MW00909723

ROBOT BOOK 2018

ISBN: 978-1-943605-03-3

Publication Date: September 19, 2017

Publisher & Author: Ahmet Tuter

Contact info: editor@roboticmagazine.com

www.RoboticMagazine.com - Robokingdom LLC

Disclaimer:

About the Author:

Author's interest in robots starts back from early school years, by building robots from kits. In 2008 he started the company Robokingdom LLC in New Jersey, and the website RoboticMagazine.Com, which became a fairly known website about robot and robotics news today, that displays news and posts from many hobbyists and companies form all around the world. In addition to almost 10 years of running of Robokingdom and RoboticMagazine.Com, he also has years of prior experience in construction industry in both east and west coast of the US, and holds a Civil Engineering degree. He speaks five languages, three of them fluently.

Since 2008, Robokingdom LLC runs several websites on robotics and construction subjects, including the news and information source website about robots, RoboticMagazine.com. This book, to a small degree follows the format of the website, by presenting the robot categories, but it also adds a lot more, by covering terms and concepts and much more.

Front Cover Photo Credits (in alphabetical order):

ELEKTROLAND - www.elektrolanddefence.com : For the tactical robot : "TMR-II TOUGH"

Istituto Italiano di Tecnologia (IIT) - www.iit.it : For the humanoid robot : "iCub"

MABI AG - Robotic - www.mabi-robotic.com : For the industrial robot arm : "Speedy 12"

Shenzen Keweitai Enterprise Development Co. Ltd. - www.keweitai.com : For UAV : "KWT - X6L"

Back Cover Photo Credits (in alphabetical order):

AERONAVICS - www.aeronavics.com : For the UAV : "AERONAVICS NAVI"

Universal Robots A/S - www.universal-robots.com : For the collaborative robotic arm : "UR5 Double"

CONTENTS

FOREWORD

Robots are increasingly becoming part of our lives. In this book, robot & robotics related topics are covered in an organized form, to serve as a learning source and general reference guide. I believe the resulting work is satisfactory, and a need has been met.

The aim of this book is to be useful to anyone who is interested or involved in robotics in some way. For example, to a beginner, the book will teach about robots, terms and concepts and give solid foundation from every angle, in plain English. To a hobbyist, or to someone whose work is related to robots, this book can serve not only as a good read, but also as a useful reference material, to be looked at again later when needed.

Terms and concepts are often explained by using other terms, which are also introduced in the book, so in a way, it is a self sufficient book. The language of the book is fairly simple, and sometimes a little technical which still includes introduced terms and still not very far from plain English. For most of the terms, related terms are also written as applicable, to facilitate better understanding.

Robot building is also included in this book, but to a small extent, to give an idea with an example that describes building a multirotor drone to a fair detail, but this book is more of a reference material, and it should serve hobbyists also in that way.

It was hard work to create this book and the most sincere efforts were spent. I dedicate this book to all hardworking and value creating readers, who would put even the fastest and best robots to shame.

I hope you find this book useful.

Sincerely,

Ahmet Tuter

Editor - RoboticMagazine.com

Robokingdom LLC

CHAPTER 1

ROBOTICS TERMS AND CONCEPTS

The list below should serve as a general reference guide and learning source. The depth of detail for each term vary anywhere from a basic definition to somewhat technical information as it was appropriate to cover, to make sure that the term's meaning and purpose is understood. Some of the items below would also belong to an electronics reference book, but they are presented here if fundamental.

Accuracy:

An important criteria for industrial robots, accuracy means, how close, or how precisely a robot can move to a desired target position in 3D space. Factors affecting accuracy include speed, robot's position within its working range, the amount of payload, links, motors, sensors. The robot's joints' assumed and actual zero positions along with its manufacturing tolerances have greatest impact on accuracy. Accuracy is a very important criteria for industrial robots because it defines how well the robot can perform a required task, in terms of accurately reaching the target. Accuracy value can be improved with external sensors such as infrared, laser distance measuring or vision system. Accuracy goes hand in hand with a robots repeatability. For comparison of both, see below:

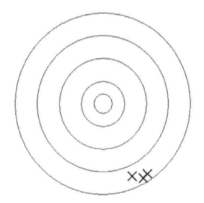 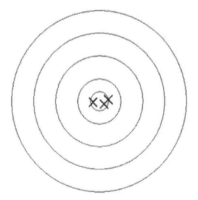 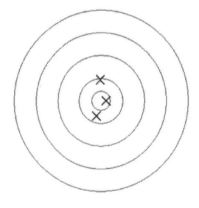

Bad accuracy, good repeatability   Good accuracy, good repeatability   Good accuracy, bad repeatability

Actuator:

An actuator is a machine component that controls the moving mechanism of parts or systems of a machine. Simply put, an actuator is used to move something. A signal and source of energy is required to activate and energize the actuator and put it to work. It basically converts any type of energy into mechanical motion either as rotational or linear. Actuators can be hydraulic, mechanical, pneumatic, thermal, magnetic and of course electric. Actuators can be considered as a subcategory of *transducers,* which is a device that converts energy from one form to another.

Actuators are used in all kinds of machines and to understand them is important. An actuator can sometimes be confused with a motor. Actuator is different from a motor in the sense that its operation is usually for a short duration and for a controlled amount of movement and is reversible. An actuator can contain a motor to provide the movement. For example, a motor can provide the necessary turning energy for a screw, where it converts turning into linear motion and this is called a linear actuator which moves a stick back and forth for any type of purpose, such as opening a door or moving a robot's leg. The "motor" part of the actuator does not need to be electrical. For example a hydraulic linear actuator works by pushing the pistons back and forth, with the energy provided to the fluids by the pump. In this example, the fluid and pump system serve as the motor. Linear actuators are the most common, and they provide push and pull motion. They can be circular or rectangular/square in cross section. Actuators are attached to a robot by mounting brackets, and can be controlled by actuator control boards, which control sensitivity, speed and stroke limits.

Apart from regular rigid actuators, there are also elastic actuators. The elastic actuator include an elastic element, (or series of elastic elements which is called a series elastic actuator) such as a spring, which gives the system tolerance to impact loads by lowering the stiffness, passive mechanical energy storage ability, low resistance of mechanical output, and therefore is very useful in force control in robotics. Elastic actuators can also amplify the power output. Having all these benefits, elastic actuators are often used in many robotic applications, such as grasping, walking robot legs, exoskeletons. As in rigid actuators, elastic actuators can also be made as rotational actuators.

Things to consider when choosing a linear actuator include force rating, stroke distance, voltage, loaded and lot loaded speed, peak power point, gear ratio, force / speed combination such as 30:1, 100:1, 200:1, linear feedback along stroke length, weight, current draw, built in limit switches.

AGV:

Acronym for *Automated Guided Vehicle.* Please visit *Drones & Robotic Vehicles > Unmanned Ground Vehicles (UGV) > Automated Guided Vehicle (AGV)* section.

Ammeter (Ampere Meter):

A device which measures electrical current in a circuit, in amperes (A). For smaller circuits, micro or milliammeters are also used, since ampere is relatively large amount of current.

Amplifier:

This is an electronic device which increases power of a signal in terms of voltage or current. It is also called electronic amplifier.

Analog Input:

Analog input signals are time varying quantities which change over time. In electronics, these are voltage or current values. Analog input values can vary and may take any value between different ranges, whereas a digital input may only take a certain value, as either 1 or 0 but nothing in between. Most electronic components are analog, such as capacitors, resistors, inductors and more. We live in an analog world, where everything can take infinite number of values. Signals can be transmitted by wires or radio frequency (RF). Many sensors convert things such as pressure and temperature into voltage. After these are measured, they can be stored in a computer by analog input and can be used for various purposes in controlling a machine. For example if a particular temperature sensor outputs 0.01 V per each degree celcius, and analog input of 0.19 V to the computer would mean 19 degrees celcius. Same can be done to measure force, where force values through analog input process can be used in many ways for robot motion control. As another example, let's say we have a power source of 8V and we connect the + and - ends to a potentiometer. The output from potentiometer will be anywhere between 0V - 8V. We can use this value in a motor driver in order to control the speed of a motor which in turn controls the speed of the wheel of a robot. Analog circuits use analog input principles. Also see *Digital Input, Electric Circuits*.

Android:

Please visit the separate chapter called *Androids / Humanoids*.

Anthropomorphic Robot:

It is a robot that is made to resemble human body as much as possible. Only some of the humanoid robots are in this category, so anthropomorphic robots are a subcategory of humanoid robots. Most humanoid robots use motors instead of linear actuators and therefore cannot truly resemble human body. At present, an anthropomorphic robot doesn't necessarily mean a more advanced robot, it just means a robot that resembles an actual human body the most. For example, even most advanced humanoid robots in the world today, are not in this subcategory, although we are starting to see examples. Currently, making a robot in this category comes with its challenges and a humanoid robot which is not in this subcategory, can still perform better, but in the future, eventually as our technology improves, the best performing humanoid robots will inevitably start to fall into this category, as only a robot which truly resembles and work like human can perform best as human.

Articulated Robot:

It is a type of industrial robot, which moves with rotary joints, therefore all movement is made possible by rotating joints but not a translating, or in other words, linear motion. The number of joints can range anywhere from 1, if it is a very simple robot, to as high as around 10, for very complicated robots. Also see *Cartesian Robot, Scara Robot, Parallel Robot*, for other types of Industrial robots. Also see *Industrial Robots* Section and the photo under that section.

Artificial Intelligence:

Please see the separate chapter called *Artificial Intelligence*.

Artificial Muscle:

Movements in robots are often achieved by motors. Artificial muscle concept is an alternative means, which aims to make robots more lifelike with higher flexibility and strength. Like real muscles, artificial muscles can reversibly contract, expand or rotate. The motion is initiated by external stimulus such as current, temperature change, voltage, pressure, change of electric field, or even by absorbing and releasing water vapor. Various materials are being tested and developed for making artificial muscles, including electrically responsive polymers, highly twisted fibers of various materials, ranging from ordinary nylon sewing thread to carbon nanotube, polymer fishing line or fibers.

According to a research led by University of Texas in Dallas and participated by scientists from universities in Australia, South Korea, Turkey, Canada and China, twisting a nylon sewing thread or fishing line, and bringing them into a helix configuration, can produce an artificial muscle with torsional strength that can spin a heavy rotor 100,000 revolutions per minute. This can also provide tensile actuation, where the artificial muscle changes in length with changing temperature and can lift 100 times more weight than a human muscle of the same length and weight. The amount (length) of actuation was also considerable. Due to anisotropic (different characteristics in different directions), material properties, even though the diameter of the thread expended only a few percent, big changes in length was observed, which also makes this a very suitable and cheap material for artificial muscles. Source: https://www.utdallas.edu/news/2016/9/27-32199_UT-Dallas-Scientists-Put-a-New-Twist-on-Artificial_story-wide.html

As the materials are subject to lose performance upon receiving any damage, even self healing materials are being worked on, just like biological muscles heal after damage. The self healing concepts can even be extended to self healing electronic circuits.

Artificial muscles is still a developing technology, however if certain performance and cost levels are reached, has the potential to be a highly disruptive emerging technology, and may find wide applications in many fields including robotics, medicine, industrial applications, smart textiles and more, due to the possibility of making them in desired sizes and configurations and their high flexibility, versatility and power to weight ratio. Artificial muscle fibers can offer so many times increase in strength with respect to natural muscle fibers of the same length, and the results confirming this are already being achieved in universities and research labs. Apart from strength, some extremely stretchy materials are also being obtained. Some of these materials can also be used in making *electronic skins*. Also see *Soft Robotics, Biobots, Electronic Skin, Haptics*.

Augmented Reality:

Augmented reality means, viewing the real world whose elements are augmented or by computer generated information in terms of graphics, sounds, haptic feedback, by turning the environment around the user into a digital interface. In other words, augmented reality can be considered as our usual everyday reality that we perceive through our 5 senses, plus, the addition to that reality with computer generated information about the items around us, therefore enhancing what we see, hear and feel or even smell, or even placing virtual objects into that environment on user's screen in some cases. This information can be anything from the properties of that object, to its coordinates or dimensions, its history, safety information, temperature, price, technical information, sound, smell or anything that may be related to that object or that place. To better visualize this, think of the screen display from an android's point of view, like in the sci-fi movies, where a lot of useful info about nearby objects or places are displayed. This information is automatically generated by computer through various sensors, and supplied to the user, such as through wearing a helmet that has screen on it, or even eyeglasses, or even a user who controls a robot remotely, that automatically brings all those info to the wearer's or user's screen. Augmented reality lies somewhere between virtual reality and the real world, but arguably closer to real world, because VR totally replaces the real world with a virtual one. Augmented reality can benefit anyone, in any job, from soldiers to doctors, taxi drivers, construction workers, astronauts, tourists, and you when walking down the street or shopping at a mall, or buying groceries. This information can be displayed most popularly as in a pair of glasses, which has been introduced to the market before, but there are still some hurdles to clear, such as the technology or privacy concerns.

Automated Guided Vehicle:

Please visit *Drones & Robotic Vehicle > Unmanned Ground Vehicles (UGV) > Automated Guided Vehicle (AGV)* section.

Automation Cell:

See *Robot Cell*. Also called *Work Cell*.

Autonomous Cars:

Also called *Driverless Cars*. Please visit *Drones & Robotic Vehicles > Unmanned Ground Vehicles (UGV) > Autonomous Cars* section.

Autonomous Navigation:

It is the process of continuously modeling static and dynamic environment of a robot, and then based on these progressing models, determining the path that the robot must take. One of the main challenges when planning the path of the robot is to predict the future states of moving objects and obstacles in dynamic environments. Inputs from sensors and cameras are continuously analyzed by navigation algorithms, in order to continuously update the robot's location as well as updating the future estimated pathways of the nearby obstacles.

This technology has seen a huge improvement since 2000s, when promoted first by DARPA Grand Challenge, and then became a widely studied and developed technology in many research institutions, universities, automobile companies and even by companies that are not primarily involved in automotive, for various reasons. Nowadays we are starting to see real world applications of this technology, which is expected to start to have a massive effect on our daily lives in the next 5-10 years, through autonomous vehicles, especially autonomous cars and delivery trucks, and autonomous mobile robots of various types and purposes. Also see *SLAM*, *Robot Mapping*, and the separate section called *Drones and Robotic Vehicles*.

Autonomous Underwater Vehicle:

See *Drones & Robotic Vehicles > Unmanned Sea Vehicles (USV) > Autonomous Underwater Vehicles (AUV)* section.

AUV:

Acronym for Autonomous Underwater Vehicle. Please visit *Drones & Robotic Vehicles > Unmanned Sea Vehicles (USV) > Autonomous Underwater Vehicles (AUV)*.

Ballbot:

This is a type robot which is designed to balance on only one spherical wheel. A ballbot keeps its balance through a feedback loop of sensors that constantly check robot's balance and make minor corrections many times per second, to keep the robot in vertical balanced position. The ballbot shown here can move in all horizontal directions, and in addition, rotate itself along vertical too, but unlike other ballbots, it uses a spherical induction motor, invented in Carnegie Mellon University, which is the only moving part of the robot shown in the picture (induction motors are nothing new but here the rotor is spherical and can turn in all directions without limit). This new type of single motor makes the maintenance process much simpler, in comparison to other ballbots using multiple motors that actuate rollers to move the ball, which is the usual case. The belts that drive the rollers in a usual ballbot eventually wear out and need to be replaced, which in turn necessitates calibration, but in this case, with only the sphere moving, that process is eliminated altogether.

A ballbot with a height of around 5'. Unlike other ballbots, which use multiple motors to actuate rollers to move the ball, this one uses a spherical induction motor as shown below. Photo Source: Carnegie Mellon University - www.cmu.edu

Ballbots are suited to working with people in human environments. They are thin robots which can move through narrow places, and can be easily pushed out of the way when necessary. The applications include carrying things, guiding people, serve as a support for elderly people or disabled people. Also see *Hobby Robots > Self Balancing Robot*.

The rotor of the spherical induction motor is a precisely machined hollow iron ball with a copper shell. Current is induced in the ball with six laminated steel stators, each with three-phase wire windings. The stators are positioned just next to the ball and are oriented slightly off vertical. The six stators generate travelling magnetic waves in the ball, causing the ball to move in the direction of the wave. The direction of the magnetic waves can be steered by altering the currents in the stators. The spherical induction motor (SIM) invented by Hollis, a research professor in Carnegie Mellon University's Robotics Institute, and Masaaki Kumagai, a professor of engineering at Tohoku Gakuin University in Tagajo, Japan. Photo & Caption credit: Carnegie Mellon University - www.cmu.edu

Battery:

Batteries are DC energy sources to power any type of robot. For most hobby robotics, NiMH and Ni-Cd batteries are used. For drones, usually LiPo batteries are used. These are all rechargeable batteries.

LiPo, which stands for lithium polymer, is the newest and the most expensive of these, which is made of carbon and highly reactive lithium, that allows it to store a lot of energy. It delivers the most power in comparison, and that is the main reason it is used mostly in flying drones, as flying requires the most amount of power, just like in nature where birds need the most protein rich prey for more energy, or jet planes needing the highest grade fuel. Despite the continuous small incremental improvements, battery technology is still the biggest bottleneck in front of drones having widespread use. Still, recent improvements are now allowing many consumer drones to stay in the air as long as 30-40 minutes, which started to make many uses possible in the recent years, and we see them increasingly in our lives. Another reason that LiPo batteries are used for drones is that they have the best weight, size to delivered power ratio, as weight is a critical component for flying drones. LiPo batteries can also be recharged in relatively shorter time. They also provide a flatter discharge rate, which means it will continue to deliver same power, even towards the end, unlike NiMH batteries, which lose energy as the run continues. Also see *Drones* chapter of this book, and look under *Specific Terms for Drones* subsection, for more information on LiPo batteries.

NiMH or Ni–MH stands for nickel–metal hydride. This can be considered as an upgrade of NiCd technology. It uses hydrogen to store energy, with nickel or another metal and therefore the name.

NiMH batteries have 1.2 V for each cell, and therefore can have multiples of this number as voltage, and they typically come in 6, 7 or 8 cells, so they will have voltages as 7.2 V, 8.4 V or 9.6 V. Despite all advantages of LiPo batteries over NiMH, one advantage of NiMH is that they are safer to use (even fire can result if LiPo is handled improperly for instance), and less care is needed to handle than LiPo batteries, they tend to endure more recharge cycles, plus they are cheaper too, so they can still be suggested for beginner level robot projects, if there is an option to choose between the two.

NiCd stands for Nickel-Cadmium, which is the oldest type of the three, which is in existence for more than 100 years. In general, they have even less capacity than NiMH batteries, although they can provide same voltage, NiCd batteries last shorter time than NiMH, which means they store less energy. Other disadvantages of NiCd batteries are that they may suffer from voltage depletion, which means that the battery holds less charge each time, if it is not completely discharged previously (NiMH also have this problem but less noticeable), they are not environmentally friendly, since cadmium is a highly toxic heavy metal, and they can endure less number of discharge-recharge cycles. The voltage depletion can be reversed if the battery is reconditioned by fully draining and charging it several times. In general, the damage from overcharging or improper storage however, is not reversible. One advantage of NiCd batteries is that they require less sophisticated chargers than NiMH and they cost less.

Battery Eliminator Circuit:
See *BEC.*

BEC:
Acronym for battery eliminator circuit, also known as voltage regulator, BEC is a circuit designed to eliminate the need for multiple batteries, by distributing power to circuits. Historically this was used for distributing power to battery driven equipment from mains electricity. For drones, it can be part of the ESC or radio control receiver or power distribution board. BEC can also adjust the voltage amount, to various parts of a robot. If the robot has more than usual servos and electric circuitry, which requires more current than a single BEC can supply, a second BEC can be used.

Biobot:
These are robots that are made from both artificial materials, and living tissues. The living muscle tissues are placed on nontoxic skeleton made of metal or polymers. The tissues can be stimulated with electricity or even light, which enable the robot to crawl or swim. These robots can be safer around humans due to soft exterior and be able to move more naturally. Much smaller of these bots made from human cells can in the future used for medical purposes too, such as targeted drug delivery to human cells. Also see *Soft Robots, Artificial Muscles, Electronic Skin.*

Biomechanics:
It is the study of mechanical laws for biological organisms, concerning their movement and structure. The mechanical study here is closely related to the mechanics that is studied in engineering disciplines. When making robots that imitate human or animal motions, biomechanics principles are

applied. In a way, it can be considered where medicine and engineering merge. As an example, walking of every person is different, and with biomechanics principles, it is possible to identify a person on a security camera, whose face is not seen, just by analyzing his or her walking.

Bionics:

In relation to robotics, bionics deals with making artificial robotic members such as feet, legs, arms and hands, in order to imitate biological members, for use by people who needs them. For robot arms that are not for medical purposes and used in industrial and consumer robotics applications please see the term *Robotic Arm*.

Bipedal Robot:

This term refers to robots that have two legs to move, which mostly mean humanoid robots, but can be of different shape too, as not all bipedal robots are humanoid. Bipedal robots have the advantage of walking on irregular surfaces or even climb stairs, therefore, as the technology develops, they can blend easier among humans as personal service robots. All humanoid/android robots are bipedal robots.

Board:

See *Microcontroller, Circuit Board, Breadboard.*

Breadboard:

A circuit for making experimental electric circuits. Solderless breadboards are very easy to use and popular among hobbyists, because different elements of the circuit are easily attached or removed which makes the experimenting process very fast. The term breadboard comes from early amateur hobbyists using actual cutting board for bread to design their radio circuits. Also see *Circuit Board.* Breadboard is also known as protoboard.

Capacitor:

A capacitor is a component of electrical circuits, which stores energy in terms of electrical field. Capacitors are one of the fundamental components of electrical circuits and can be made in a large number of ways and therefore many different types of capacitors with different names and construction are in use today.

Cartesian Robot:

A type of *industrial robot*, which is linearly actuated on three perpendicular axes as x, y, z, but it does not have any freedom to rotate. Also see *Articulated Robot, Scara Robot, Parallel Robot,* for other types of Industrial Robots in terms of how they move. Also visit *Industrial Robots* Section.

Cell:

See *Robot Cell.*

Central Pattern Generator:

It is a neural network, which can produce rthymic output, such as walking.

Chip:

See *Integrated Circuit*.

Circuit:

See *Electric Circuit, Circuit Board*.

Circuit Board:

Circuit boards connect electrical components by conducting (i.e. copper) pathways, and it also serves as a framework for the circuit to attach. Circuit boards are also called PCB, printed circuit board. PCBs can easily be custom made based on someone's need for any electronics or robotics project. A PCB can be single or double sided or multilayered.

To design a PCB, first, what circuit will be built must be known, then schematics must be drawn, and finally this is transferred into a circuit board layout. There are many PCB design software available. By using the software, the PCB design including layout is finalized and it is made ready for manufacturing. A standard file type format used for this, is called a Gerber Format, developed by Gerber Systems Corp, which was later acquired by Ucamco Company https://www.ucamco.com/. It

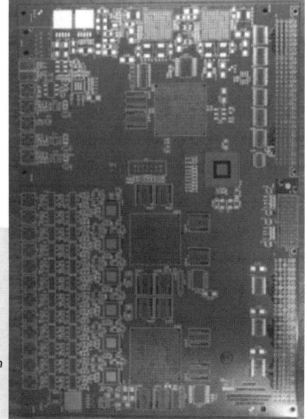

is a file that shows how each layer of a circuit looks like. So for instance, for a two layer circuit board, there will be files such as top and bottom, silk top and bottom, solder mask, and drilling layer. This is the information needed by a machine in order to create that layer of the circuit board. Both high end and beginner level circuit board making programs use Gerber files.

When designing a PCB, size of the board, drill holes and traces are some of the main limiting factors for layout. The PCB manufacturers will have guidelines as a checklist, to help in the process. The size of the board is the first to be considered, as the board must fit

A circuit board example. This is a 6 layer high temperature FR-4 printed circuit board that is 0.062" thick, has green solder mask, white silkscreen (also known as legend), and a Electroless Nickel Immersion Gold (ENIG) surface finish. Photo & Caption Credit: Custom Circuit Boards www.customcircuitboards.com

somewhere in the end product.

Trace line widths must be as thin as possible to save space but not so thin as thinner traces are more vulnerable to damage when soldering, plus there is a current it must carry, which requires a certain minimum thickness. All traces must be routed. Power traces must be thicker than signal traces as they carry more current and they should not change characteristic over its length. Heavy components such as connectors, transformers, coils must be placed near the edge for easier connection. Proper labeling must be printed on the board for all components, to solder accurately and easily troubleshoot later. The design must be checked for errors

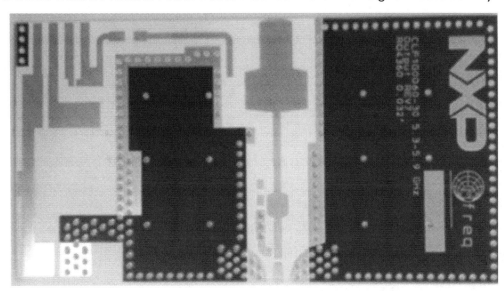

A circuit board example. This is a 2 layer Rogers 4360 bare printed circuit board that is 0.031" thick, has a blue solder mask, white silkscreen, and a Omikron surface finish. Boards built with Rogers 4360 material are typically needed for high frequency applications (antennas, etc). Photo & Caption Credit: Custom Circuit Boards - www.customcircuitboards.com

before sent to manufacturing, such as checking for unrouted nets, board and schematic layout differences.

Cleaning Robot:
See *Vacuum Cleaner Robots* under *Domestic/Consumer Robots* chapter.

Clock Speed:
It is the speed which a microprocessor executes operations. Also called clock rate. Every computer contains an internal clock that regulates and synchronizes the operations of its components.

Co-Bot:
Short form for collaborative robots. These robots are made to work alongside humans, in order to assist humans in many different ways, such as performing repetitive tasks which would be too dull for humans to do or lifting heavy items that would put strain on a human worker's body. Working alongside humans necessitate safety measures to prevent injuries to human workers, such as limiting the force of the robot and stop it, as soon as it senses an external unexpected force or impact,

through its force and torque sensors, which can be located on the arm and the base, and also covering the outer shell of the robot with soft material to reduce any damage. (This is not the case for usual industrial robots, where they are designed to work on assembly lines or workcells which are not supposed to include humans.) Because of the efficiencies involved, just like the industrial robots, many firms are eager to adapt this technology. Collaborative robots have higher level of software, visual recognition and manipulation abilities in comparison to usual industrial robots, because in the collaborative robots case, certain level of autonomy and decision making is needed in order to keep up with the humans, as opposed to industrial robots, which only adhere to the tasks they are programmed for, in unchanging environments and situations and in heavy industrial settings only. Training of these robots are also as easy as showing the tasks to be done manually. Collaborative robot industry is developing fast and will see big growth in the next decade, as there are a huge number

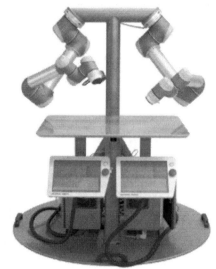

A Co-Bot example. Photo Credit: Universal Robots A/S www.universal-robots.com

of possible applications and many companies are interested in developing and commercializing these types of robots.

Cognitive Computing:

Cognitive computing can be described as the field of artificial intelligence, where the intelligence is achieved with humanlike traits such as hearing or seeing and trying to process the information similar to humans. In other words, cognitive computing seeks to achieve human thought process the same way humans do, in a computer. Many fields come into play for cognitive computing such as machine learning, speech and visual recognition, pattern recognition, natural language processing, neuroscience. The cognitive computing system must be able to adapt to new situations, understand context and interact accordingly. The machine learning algorithms used for cognitive computing can work on a large amount of data (such as by data mining) which can give it a huge edge over humans, in terms of the quantity and speed of information is processed. The systems can not only process information but they can also improve the way they understand patterns and adapt themselves to previously unanticipated new situations. Today cognitive computing is used in expert systems. Also see *Deep Learning, Artificial Intelligence.*

Collaborative Robots:

See Co-Bot.

Comparator:

An electronic device that compares two voltages or currents and gives output, as to which one is larger.

Component Forces (Force Components):

The components of a force vector in two or three perpendicular axes. Expressing a force in its components, makes static and dynamic analysis easier, because the components in all axes can be summed up, in order to determine the resultant forces, to find previously unknown reaction forces. In the figure below, you are able to see a force F, and its x and y components, Fx and Fy. Same principle would apply if this was in 3D, with the addition of z , but it is not shown here for clarity. Here the magnitude of force F can be expressed as, $F^2 = F_x^2 + F_y^2$, just like calculating the hypothenus of a triangle. Finding components makes calculations easy because for example here if we wanted to add this force F with another force, which may have different direction and magnitude, we simply would resolve both into their x and y components, add the components together, and then find the resultant force magnitude from the formula above. The ratio of the components of the resultant force will also give us the direction of this force, such as a new angle alpha, by calculating the angle from tangent. Keep in mind that force is a vector, which means, it has both magnitude and direction. Understanding forces and vectors are essential to calculate movements of robots. Also see *Force, Free Body Diagram*.

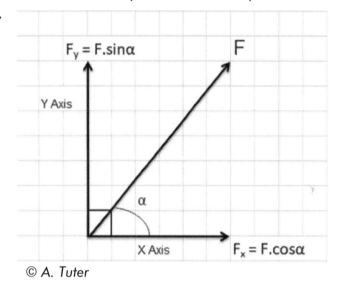

© A. Tuter

Computer Architecture:

In construction industry, architecture means designing buildings or other structures, by using available materials such as concrete, steel, brick, glass etc... and doing this in a cost effective manner, at the same time trying to achieve high quality, safety, reliability, efficiency and functionality. That is why it is often considered an art, in addition to being science. Architecture in computer industry is similar concept. Computer architecture seeks to build computers, game consoles, phones and other hardware in a cost effective, efficient manner achieving functionality, high quality, reliability, by using available technologies (similar to materials in construction) such as logic gates, memory technologies, circuit technologies, storage technologies. Of course, technologies evolve at a much greater speed in computer industry and therefore computer architecture is evolving rapidly. For a beginner level hobbyist who wants to build robots, it is good idea to have at least a very basic understanding of computer architecture and embedded computing. Also see *Embedded Computing*.

Computer Vision:

Computer vision deals with how a computer receives, reconstructs, recognizes (interprets) and processes or understands visual data, in order to produce useful meanings out of these data. It encompasses many fields, such as computer science, mathematics, electrical engineering, cognitive science, and even medicine. For example, an autonomous car camera views the street, produces meaningful data from these visual scans, and then intelligent decisions are made based on this process to guide the car. When computers and humans are compared in terms of visual processing, there are things that humans do better, such as recognizing a friend's face, or a pattern, but many other tasks are better done by computers, such as calculating the exact speed of an approaching car, processing an image accurately that would be too fast for a human eye, remembering and not forgetting any image that was seen and not being subject to any illusions. Therefore, it can be said that the ultimate goal of computer vision is to model, replicate and finally exceed human visual recognition capabilities, using computers and hardware. Computer vision is used in many other fields in addition to robotics. Examples include, surveillance, medical image analysis, agriculture, augmented reality, process control, remote sensing, geoscience, transportation, face recognition, pollution monitoring and more.

Computer vision can often be confused with machine vision, as they are closely related. Machine vision deals with analyzing and inspecting images in industrial settings, such as for vision guided robots, where computer vision has broader sense and applications as discussed, so it is the parent category of machine vision, but a subcategory of machine learning. Also see *Augmented Reality, Cognitive Computing, Machine Learning, Object Recognition,* and the separate section of *Artificial Intelligence.*

Concurrent Forces:

The forces which act towards a common point, through their line of action. It is a term used in statics and dynamics. When studying robot dynamics, such as the dynamics of a walking robot, the forces are represented by force vectors. If all the forces are concurrent, which means their line of action intersect at a single point, there will be no resultant moment. If at least one force is not concurrent with other(s), then there will be a resulting moment, in other words, a turning effect to the robot or the robot component in focus.

In the figure below, forces F1, F2, F3 and F4 are concurrent, which means, their lines of action (it is shown as a dashed line extending from both ends of force vector, but not shown here for clarity) intersect at a single point, which is point 1 here. So they all act on point 1, even F3, because, a vector could be slid anywhere along its line of action, as long as its magnitude and direction is not touched. Same is true for F5, F6, F7 forces, which intersect at point 2. If we had a robot body, which had only F1, F2, F3 and F4 forces acting on it, there would be no turning effect on the body, because all forces intersect at point 1.

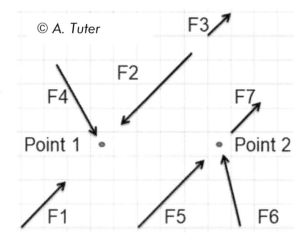

However, in addition to these four, if you were to add any one of F5, F6 or F7, then, there would be a turning effect (moment), because the added force F5 or F6 or F7, does not act on point 1, and would cause a turning effect around point 1 (or equivalently, the resultant of F1, F2 F3 and F4 would cause a turning effect around point 2). Also see *Force, Component Forces, Zero Moment Point, Free Body Diagram.*

Coupled Force (Force Couple):

A force couple must satisfy all of the following conditions: Two forces that are equal in magnitude but in opposite direction, which are not concurrent , but parallel. This produces a moment, which is a turning effect on a body.

In the figure below, forces F1 and F2 make up a force couple. Their lines of action are parallel, so they produce a moment equal to

Moment = force 1 x d = force 2 x d

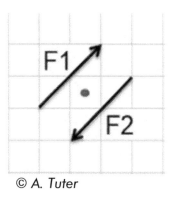

© A. Tuter

Here d is the distance between the parallel lines of actions of forces.

Note that the result would have the same magnitude if we calculated moment as

Moment = force 1 x d/2 + force 2 x d/2

Also see Force, Concurrent Forces, Moment, Free Body Diagram.

CPG:

Acronym for *Central Pattern Generator.*

Current:

Current is a physical quantity, that represents the flow of electric charge Q, at a point in a conduit. It can be carried by electrons in a wire or ions in an electrolyte or by both, as in a plasma. The unit of measure for electric charge is ampere. It is related to the charge Q, with the following equation: I=Q/t, where I is the current, Q is the charge, and t is time. Here charge Q is in coulombs and t is in seconds. With this relation, it can be understood that current is a rate quantity, that is, the rate of charge that passes a point in a circuit in a given time. A current of 1 ampere means, a charge of 1 coulomb passes through a conduit in 1 second.

Electric current causes heating in a conductor and creates a magnetic field. Current is directly proportional to the voltage, by a ratio called the resistance. This is expressed by Ohm's Law: I=V/R, where I is the current, V is the voltage and R is the resistance in ohms. If the current flows in one direction it is called a direct current (DC). If the direction of current periodically reverses direction, it is called alternating current (AC). Batteries provide direct current, whereas the outlets we see in our houses provide alternating current. Current is measured by a device called *Ammeter*.

Cyborg:

This is the short form for cybernetic organism which represents a being with both biological and biomechatronic parts. Popularly, a cyborg mostly means an android robot, which is a robot in human form and looks identical to a human. Also see *Biobots, Androids / Humanoids.*

Data Mining:

Data mining is the technique of searching a data set and finding meaningful and relevant patters or features, which was previously unknown and cannot be detected by humans. This is made possible by processing power of computers combined with necessary algorithms. Also see *Artificial Intelligence* section.

Deep Learning:

This is a subfield of *Machine Learning*, please see that term.

Deep Neural Network:

This is an advanced, layered type of neural network. Please also see *Neural Network*.

Degrees of Freedom:

This term actually is used in many different fields, such as mechanics, structural engineering, robotics, statics, physics, chemistry and probably more. In broad terms, it is the number of parameters in a system that can vary independently. For example in statics, it is the number of parameters obtained from calculation that are free to vary. For robotics and structural purposes, it means along or around how many axes in total, a mechanical body is free to move or rotate. The higher the number of degree of freedom, the more complex a robot gets but it also gains more ability to make more variety of motions. Think about your arm, it can move in any direction, and also rotate, not only from one but multiple joints, which means it has a great number of DOF, whereas a rear tire for a car has only 1 degree of freedom, which is rotation about only 1 axis. A front tire however, has 2 DOF, as it can rotate along 2 axis, by also turning when the steering wheel is turned. A sliding door has 1 degree of freedom, only along the axis which it slides, or a rotating door also has 1, as it turns only around 1 axis. You head has 3 degrees of freedom, as you can rotate your head around all three axes, but it cannot make translation movement in any of these axes. When you are standing on the ground, you have 5 degrees of freedom, as you can rotate your body around any axes (at least to a point), and you can move back and forth by walking, but you cannot move upwards (if we assume you will not be jumping). A structural beam, which has a pinned connection on one end, and a roller connection on other end, has 3 degrees of freedom, as it can only rotate at the pin connection, plus, it can rotate <u>and</u> move (translating move) where the roller support is, for a total DOF of 3. A bird or a drone, when flying, has 6 degrees of freedom, as it can rotate and move around and along on all 3 axes (dimensions). That is why gyroscopes are made for 6 axes for example. A 6 degree of freedom is basically as free as a <u>single</u> object can get, as it means it floats in the air. Understanding the concept of DOF is important in mechanics and therefore robotics, hence a lot of examples given here. An industrial robot for example, may have many degrees of freedom, in order to gain ability to make moves required for production. A humanoid robot, may have a total DOF in the range of as high as 50s, when all DOF for different joints are added (do not be confused, we said that a single body can have at most 6 DOF, but the 50s here means all DOFs from different members of a robot, which are all single members separated by joints, are added to each other to reach to a sum). The final position of a robot arm is calculated based on the cumulative movements of each joint, whether translation or rotation. DOF is one of the most important characteristics to know about a robot.

Depth Sensing:

See visual 3D depth sensor under *Sensors* chapter.

Digital Input:

Digital input values can only be a value or zero but nothing in between. This is unlike analog input, where any intermediate value is also possible. For example, a proximity sensor with an output of 12V, will output a voltage of 12V upon detecting a nearby object. This output is digital and goes to a control board, which in turn energizes a motor with that value, in order to provide motion. Most of our computers, logic units and microprocessors are digital, but the word we live in is an analog world, with infinite values in between. Digital circuits use the digital input principles. Also see *Analog Input, Electric Circuits*.

Digitizing:

Digitizing is representing an object in terms of numbers, that can be processed by a computer. It also means to measure an area or perimeter by representing its boundaries numerically.

Diode:

It is a mechanism that allows electric current to pass in one direction, while blocking flow in the opposite direction, therefore it can act as a valve in a circuit.

Direct Drive Mechanism:

It is taking the movement directly from a motor, without any intermediate system such as a gearbox. The obvious advantage is the efficiency, due to avoiding efficiency loss in a gearbox for example. This also translates to longer lifetime, reduced noise, higher torque, faster and more precise positioning. The disadvantages are the need of special motors, higher torque producing motors, due to absence of a gearbox, and more precise control mechanisms to manage direct transfer and accuracy. More precise control can be done by computerized motor control and rotor position sensors, which makes this much more complex motor control mechanism. Fans, hard drives, sewing machines, washing machines, CD drives, telescope mounts are examples that use direct drive mechanism all with different speeds and precision.

DNN:

Acronym for *Deep Neural Network*.

DOF:

Acronym for *Degrees of Freedom*.

Domestic Robot:

Please visit the separate chapter called *Domestic / Consumer / Service Robots*.

Dot Matrix Display:

It is a device to display simple information with low resolution, such as on clocks, with simple led lighting working together to display numbers or letters.

Driverless Car:

Also called autonomous cars. Please visit *Drones & Robotic Vehicles > Unmanned Ground Vehicles* section.

Drone:

Please visit the separate chapter called *Drones and Robotic Vehicles*.

Dynamic Balance Control:

This study must be done for legged robots, such as bipedal (such as a humanoid) or quadruped, but mostly critical for bipedals, where the robot must be able to keep its line of gravity or center of mass within its base, while moving. Movement is the keyword here. This study is primarily done in medicine, for patients with walking difficulties. Note that it does not matter if we said here center of mass or line of gravity, as gravity on an object is assumed to act on center of mass. To keep dynamic balance, there must be sufficient power in muscles supporting the balance, sensing information regarding posture, movement, and environment, and acting on it with correct algorithms and motion control in order to keep balance is needed. Dynamic balance control, which involves movement, is more complicated process than keeping balance while at rest. Also see *Static Balance Control, Zero Moment Point*.

Electric Circuit:

Electric circuit is a path where electrons flow by the push of a voltage source. it uses two types of current, AC and DC, as explained under the term *current*. A discontinuity in a circuit, regardless of where it is, will stop the electron flow and the circuit from functioning, which is called an open circuit, otherwise it is a closed circuit while working. An electric circuit can be classified into two, as series or parallel based on how the current is routed.

In a series circuit, the same current flows through all components, and therefore when a component fails the circuit will stop to function. A good example to this is a string of Christmas lights. In a series circuit, total voltage equals the sum of the voltage of all components, such as $V_T = V_1 + V_2 + V_3$ ... and also the total resistance is the sum of all resistances, such as $R_T = R_1 + R_2 + R_3$ ... In a parallel circuit, the current flows through different branches with different components and therefore resistances. If a branch fails, the current can still flow through other branches. A good example to a parallel circuit is the electrical system of a building. In a parallel circuit, the total current is the sum of all currents in separate branches, so the equation becomes $I_T = I_1 + I_2 + I_3$ ... and from here, V and

R can be substituted from the general equation of V = I x R, which was also mentioned under the term *current*. Circuits that are used in everyday life electronics and robotics are complex combinations series and parallel circuits.

A short circuit happens, when a low resistance path is unintentionally formed, that bypasses a part of a circuit, which can cause a large amount of current to flow instantly and this can generate heat and cause fire. Therefore there are safety measures in a circuit, such as fuses or circuit breakers, to stop this unwanted current flow, by instantly and automatically opening the circuit. An electric circuit can also be classified into two types, analog circuit and digital circuit based on the signals they use. analog and digital circuits work with the principles as explained under the term *analog input* and *digital input*. Also see *Circuit Board, Current, Analog Input, Digital Input*.

Electromagnet:

It is a magnet that is only active when there is an electric current, which produces a magnetic field and can act as a real magnet. When the current is turned off, an electromagnet stops to function.

Electromotive Force:

The potential difference (*voltage*) that causes *electric current,* which is created by an electrical energy source such as a dynamo or battery. It is measured in volts.

Electronic Skin:

Also called synthetic skin, it is a thin electronic material component that is made to work like human skin in a number of ways such as sensing heat or cold, pressure, healing itself, being able to flex and more. Skins made of very light, comparable to feather, and flexible materials are made, and use thin film transistors for circuitry, which can be printed on plastic films. In some cases even memory components are present on the skin acting like a computer. These films, which are easy to handle and cheap, are very suitable for electronic skins. They can be stretched and folded, while the printed electronics on them still working. The first application that comes to the mind is of course in biomedical field. These can cling to human skin, bend and be stretched, as the person carrying it moves his or her body. It can be used for health monitoring systems and wearable medical instruments. For example they can sense symptoms on your body and deliver drugs when needed, through the skin, in addition to other functions. In the future, these can serve as synthetic skin for androids or other type of robots.

Electronic Speed Controller:

Electronic speed controllers, (ESCs) manage the speed and direction of motors. ESCs are especially used for RC devices including multirotors or remote controlled cars. Each motor has one ESC that supply power and current about how fast it should turn. This ultimately means ESCs control in what direction, and how fast the drone moves, how it turns or accelerates and its direction. ESC may include *BEC (battery eliminator circuit)* which distributes power to more than one circuit and LVC (low voltage cutoff), in cases where the voltage drops below a certain level. ESCs draw their power from the power distribution board and then transfer this power to motors. Therefore ESC must be able to provide enough current to the motors that the motors can handle. By varying the timing and the amount of current, the ESCs control the speed and direction of the motors. Most ESCs that are used for hobby grade drones are usually up to 30 amps. Also the refresh rate, which means, how many times per second the ESC checks for instructions from the flight controller, in other words, the speed of the ESC, can make difference in how well the drone is controlled. ESC converts the DC current of the battery into three phase AC signals for the brushless DC motors. These continuous signals produce continuous changes in magnetic field of the motor and create rotation. The flight controller sends a signal to the ESC telling it how fast to turn motor. There is a circuit in the ESC that converts

the signal from the controller into the much more powerful three phase signals need by the motor. In the figure below, the flight controller and power distribution board are not shown for clarity. Also see *Motor Driver & Controller, Servo Driver & Controller.*

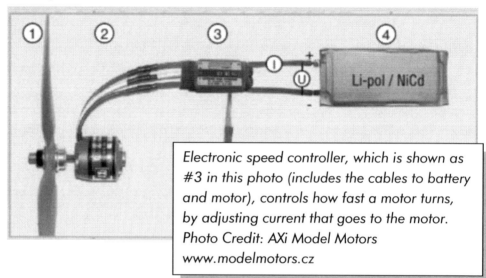

Electronic speed controller, which is shown as #3 in this photo (includes the cables to battery and motor), controls how fast a motor turns, by adjusting current that goes to the motor.
Photo Credit: AXi Model Motors
www.modelmotors.cz

Embedded Computing (Embedded Systems):

Also called embedded systems, embedded computing can be described as a computer part of a larger system, that uses its own microprocessor. Recently this subject has become an engineering discipline by itself and became critical for the design of any electronic product. Products that range from space shuttles to simplest of home appliances all use embedded computers. Embedded systems make electronic products work more efficiently, by handling calculations for a specific component of a larger system, which would be far too complex algorithms to manage with a hardwired logic without embedded systems. Software architecture, hardware platform, and power consumption are important considerations when creating embedded computers, like any computer. For a robot hobbyist, it is recommended to have a basic understanding of embedded systems and computer architecture. Also see *Computer Architecture.*

Encoder / Decoder:

An encoder is a device that converts information from one format to another. It can also be a circuit, a software program or even a person for the purpose of compression or transmission. Decoder is the entity that decodes the received info into original data.

End Effector:

The component of a robotic arm that is at the tip of the arm, which is the actual part to manipulate objects. For example, your hand is the end effector of your arm. There are different types of end effectors as grippers, pins, robotic hands, paint or welding guns, vacuums and more. Also see *Robotic Arm*.

ESC:

See *Electronic Speed Controller*.

Exoskeleton:

Exoskeletons are the robotic wearable devices for amplifying strength or for helping patients with movement difficulties. Please see the separate section called *Medical Robots*.

Expert Systems:

The systems that assist human decision making, by utilizing the processing power of computers in order to locate relevant information among large amounts of data are called expert systems. Also see *Machine Learning, Artificial Intelligence*.

FET:

See *Field Effect Transistor*.

Field Effect Transistor:

It is a type of transistor that uses small voltage to control current, commonly used for amplifying weak signals whether analog or digital signals. FET has two subcategories such as *junction FET (JFET) and metal-oxide- semiconductor FET (MOSFET)*.

Force:

A force is a vector, which means it has both magnitude and direction, which has the unit of newtons. They are the effects of push or pull on a body. A force, if unbalanced, will create an acceleration on the body, which is directly proportional to the force, and inversely proportional to its mass, from Newton's equation $F=ma$, where F is the force in newtons, m is the mass in kg and a is acceleration in m/s^2. If the force is balanced, then it means it has created a reaction of equal magnitude and exactly opposite direction. For a body to be at rest, the resultant force (the force that is the result of adding all separately acting forces by their magnitude <u>and</u> direction) acting on that body must be equal to zero, which means, the resultant magnitudes in all directions, after added with each other, must be zero, otherwise that object will be in acceleration, under a net force, from $F=ma$.

In order to determine the unknown reactions that support a structure, a *free body diagram* of the structure is drawn, all forces are represented by vectors in proper directions, and then the forces in all directions are summed up, so that the sum will be equal to zero. This will give the resultant reaction, provided that there are not too many redundant reactions. If there are, more advanced methods are used such as virtual work, displacement or other methods, which are out of our scope. Same concept applies to moments, which are also vectors, which is the turning effect of a force, which is given by M=F.d, where M is the moment, in newton-meter, F is force in newtons, and d is distance in meter. Moments are shown by the axis around which they effect. Again, for an object to be at rest, the resultant moments on a body must also be equal to zero. Studying forces and moments on static or dynamic bodies, such as structures or machines, form the basis of statics and dynamics. Forces and moments must be understood very clearly and represented properly, in order to build robots and study robot motion. For example when you buy a servo, its torque rating is given, which is the equivalent of moment, or in order to build a robot arm that can move objects of certain weight in a certain way, you must study forces acting on it, or when building a walking robot leg, all forces that will cause the dynamic walking effect must be considered. Also see *Component Forces, Coupled Force, Concurrent Force, Zero Moment Point, Free body Diagram.*

Free Body Diagram:

It is the graphical representation of all forces and moments that act on a rigid body. After a free body diagram showing all forces and moments are drawn, it can be analyzed for finding the reaction forces on the object and for static stability or dynamic movements. Let's illustrate with a simple example:

Example:

In the figure below, a beam is shown, as in Figure 1, with a force acting on it, and has two supports at its ends. For clarity, a 2 dimensional figure is presented, although in real life, robotics may often involve 3 d problems.

Let's assume the support at point A is a pin support and the support at point C is a roller support. In our example the magnitude of the force is given as 20 N, and it makes 30 degrees angle with the horizontal x axis, and it acts 4 meters from point A and 6 meters from point C, as shown in figure.

© A. Tuter

Figure 1   Supports and Applied Forces

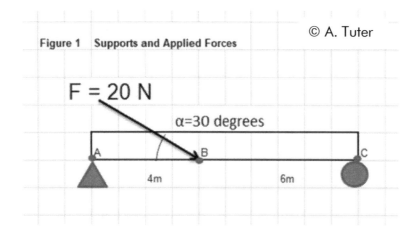

In order for this body to be at rest, the resultant force and moment on this body must be zero, which in turn means, the resultant forces and moments along all axes must be zero. This is another way of saying, that the reaction forces and moments that are created by the supports, in reaction to the applied force, must be in such a way that all the forces and moments cancel out, and the body will be at rest. So total forces in all axes, Fx, Fy must be zero, plus the moment Mz along z axis must be zero. If we were solving 3D problem, we would have additionally Fz = 0, Mx = 0 and My = 0, creating 6 equilibrium equations.

To determine the reaction forces, a free body diagram must be drawn. See Figure 2 below. This is a graphical representation of the object, plus all forces and moments acting on it, in proper locations. In our example there is no external moment applied to our object.

To begin drawing the free body diagram, all of the forces must be separated into their components, and, the supports must be removed and represented by reaction forces.

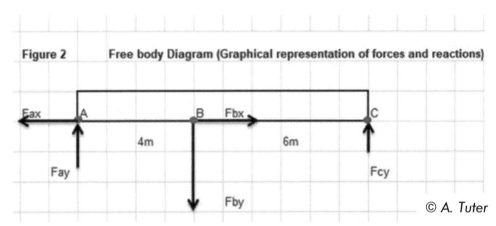

First, let's separate the acting force at B, into its components. We write them as Fbx and Fby.

Now we must show the points of supports as reaction forces.

Figure 2   Free body Diagram (Graphical representation of forces and reactions)

© A. Tuter

If you consider the pin support at point A, it cannot resist any turning effect, so there is no moment reaction there, but it can resist any movement in horizontal or vertical direction. Therefore, this resistance must cause reaction forces. We represent these reaction forces as Fax and Fay.

Similarly, if you consider the roller support at point C, this rolling support cannot resist any moment either, so there cannot be any moment reaction there. In addition, it also cannot resist any force along horizontal x direction, so it also cannot produce any reaction in x direction. Therefore, all it can create is a reaction force in vertical y axis direction, which we represent by Fcy.

To find the reaction forces, we must again remember that all forces and moments on the body must be zero, in order for the body to be at rest. Now we have our free body diagram as in Figure 2, and let's look at it.

First let's find the components of force acting on B.

From simple trigonometry, Fbx must be

Fbx = F . cosα = 20 x 0.87 = 17.3 N

and Fby must be

Fby = F . sinα = 20 x 0.5 = 10 N

Again, for the body to be at rest, the force components on all axes must be zero.

Therefore,

$Fx_{Total} = 0$ and $Fy_{Total} = 0$

For forces in x direction, we have only Fbx and Fax.

Therefore Fbx + Fax = 0

We already know Fbx, as we found above. So,

Fax + 17.3 = 0

Fax = - 17.3 N

Note that the sign is minus, which indicates it is in opposite direction to what we considered as plus. It doesn't matter which direction you choose as plus (here we had taken direction to the right as plus, so this means horizontal reaction at A is acting towards left, since it came as minus)

Now let's try to find Fay and Fcy. All forces on y direction must also be equal to zero. So,

Fay + Fby + Fcy = 0

We already know Fby, which is 10 N.

so we get

Fay + 10 + Fcy = 0

We have only one equation but two unknowns. We must have one more equation.

That additional equation comes from moment. Remember that we said for a body to be at rest, not only forces, but moments must also cancel out and the resultant moment must be zero. So, if we were to take moments of forces with respect to any desired point on this beam, and make it equal to zero, we will have our additional equation.

Let's consider the moment with respect to point C for simplicity. Remember, we can choose anywhere on the beam, since no part of beam is turning, and we will get same results.

With respect to point C, the force at C does not create a turning effect. Because it acts directly on C. So we can leave that one out of our new equation. Furthermore, we can also leave out Fax and Fbx, because their lines of action, pass through point C, and they have no distance with point C, so they also cannot create any moment with respect to C. So we leave those out too. Then, the only forces that can cause moment around point C, is, Fay and Fby and their total moment must add to zero. So let's write our new equation:

Fay . 10 + Fby . 6 = 0

Remember that for moment we multiplied Fay by its distance to C, which is 10, and Fby by 6 meters.

We already knew Fby, which was 10 N.

so,

Fay.10 + 10x6 = 0

This gives us

Fay = -6 N

Note that it is minus, which indicates that it is in opposite direction to where we took as plus, which was downwards, so it means Fay is upwards. If we put this in the other equation we solve Fcy as:

-6 + 10 + Fcy = 0,

which makes

Fcy = -4 N and which means it is upwards.

If you missed some of the details above no problem, just compare figure 1 with 3 now. And you will see the big picture.

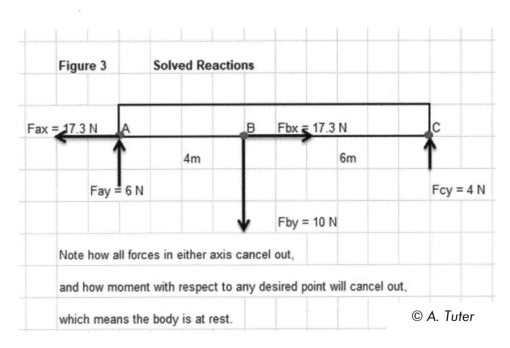

Figure 3   **Solved Reactions**

Fax = 17.3 N   A   B   Fbx = 17.3 N   C

4m   6m

Fay = 6 N   Fcy = 4 N

Fby = 10 N

Note how all forces in either axis cancel out,

and how moment with respect to any desired point will cancel out,

which means the body is at rest.   © A. Tuter

Here, if you understood this example, (although it is of course not necessary to continue to read this book), you have learned the very basics of statics and structural engineering, which was important to understand forces. In many cases we may have more reaction forces than we can solve with only the equilibrium equations, which are called redundant reactions, and in those cases, which is called the indeterminate case, as opposed to a determinate case here, other more advanced methods are used to solve reactions, which are out of scope of this book. Also see *Force, Zero Moment Point*.

Full Actuation:

This is a motion system where the number of axes that can move or rotate with motors or actuators is equal to the total DOF. See *Underactuation* for more on this. Also see *Degrees of Freedom*.

Fuzzy Logic:

This is a logic system that a variable can be anywhere between 0 and 1 however for a binary logic a variable can only be 0 or 1.

Gait Modulation:

It is the control of movement of legs of a legged robot so that it runs and walks properly. Also see *Dynamic Balance Control, Underactuation, Quadruped Robots*.

Gripper:

Robot grippers are used to grip and manipulate objects. They can be claw, vacuum or clamps. Some grippers are used in conjunction with cameras, in order to position the gripper better, and sensors in order to sense the forces. There are very eloquent gripper products today that can imitate human hand movements. The gripping mechanism can be based on squeezing the object, using suction, adhesive forces or penetrating the object. Also see *Android Robots, Industrial Robots* sections.

⊕ Gripper

This 2-Finger Gripper can be integrated for use in various kinds of automation cells in a factory. It can handle parts of different sizes and shapes.

Photo Credit: Robotiq
www.robotiq.com

Guided Robots:

See *Vision Guided Robots.*

Haptics:

In general, haptics can be defined as a science dealing with the interaction between two entities by touching, creating and transmitting the sense of touch. These entities can be either from the real world by a teleoperated system or a virtual environment constructed by computer visualization tools. For computers it means adding touch sensitivity to an application so that it can be controlled by touching. Although a relatively new technology, haptics has found many applications in many fields, such as in teleoperation, humanoid robots, exoskeletons and rehabilitative devices, entertainment industry as in computer gaming, virtual reality, medical industries such as in robot assisted surgeries, to compliment the positioning algorithms.

Haptics is not only about sensing inputs but also about sending feedback to touch sense of the user. For example, it helps medical surgeons, by transmitting the feel of touch of the surgical robot to surgeon, or a box match with someone who is at the other end of the world is possible, by receiving inputs and sending outputs through gloves and helmets which are equipped with haptic sensors.

Reverse electrovibration is a sub-area of haptics, which focuses on transmitting data to mimic human touch for the surfaces that the haptic equipment touches. Imitating of touch sense, and transmitting sense of touch is more difficult than transmitting virtual or audio through a computer. For visual, there is the computer screen, which transmits data to eyes. For audio, there is the audio speaker, which transmits sounds to ear.

To transmit feelings of smoothness, texture, temperature, geometric properties, force to human hand however, which is the primary organ to experience the world through touch, is relatively complex. For example, to transmit the feeling of softness versus hardness and solidity versus fluidity of a surface, a spring mass damper system can be used. (A mass spring damper system is a representation where a mass is pulled by a force, and this pull is resisted by a spring and a damper together). A harder surface will be represented by a stronger spring, with higher spring constant, and a more solid surface will be represented with a higher damper constant. Elastic surfaces can be represented by vibration force, which can increase or decrease gradually, at different levels, to represent different degrees of elastic surface.

Human hand is made of 27 bones, and has 22 DOF, which gives it a huge dexterity and imitating these movements and the senses of the skin that covers all this complex mechanism, which also has a great number of nerves, has is a very difficult tasks in robotics. To convey sense of touch, also soft robotics principles are used, which is a new field of robotics. Also see *Robotic Hand, Soft Robotics, Synthetic Skin.*

Heat Sink:

Design of electronics also includes dealing with heating problem, as electronics components heat during operation. Motor drivers, CPUS, motors, batteries, voltage regulators, diodes, integrated circuits and all electronic components get hot during operation. Warming of all these components affect a robot's performance negatively or may even cause damage. For example when the CPU heats up, it slows down because of an effect called thermal throttling. Although these components can dissipate heat on their own to a limited degree, this is not enough. Heat sink is a component that keeps electronic components from overheating, by absorbing excessive heat. The heat is transmitted from the electronic components to heat sink, which has fins that provide large surface area for heat dissipation. Sometimes to increase the surface area that dissipates heat even more, the fins can also be cross cut. Heat sinks can also be cut to required sizes to fit anywhere in a robot or a drone. Heat sinks are connected to these components by heat sink adhesives or thermal tapes which are good heat conducting materials but it should be used only at the minimum necessary amount. A good bond is necessary for better heat transfer.

Heat sinks for robotics projects are usually made of aluminum, which is a great material for heat conductivity, in other words, has a high heat transfer coefficient. Another reason that aluminum is very suitable material is that it can be shaped very easily with minimum effort or tools and equipment. Copper for example, although it can absorb and conduct heat even better than aluminum, with a higher heat transfer coefficient, and it also has more biofouling resistance, it weighs three times more (denser) and relatively more difficult to machine and therefore it is used less. There are other materials with very high heat transfer coefficients, but usually they are not suitable to use due to high costs. The unit of heat transfer coefficient is: $W/m^2.C$, which is Watts/squaremeter.Celcius

Heat sinks can be classified as active and passive. Active heat sinks need power source and moving parts, such as a fan, where passive heat sinks has no moving parts, and they transmit hear through convection, such as the fins of an aluminum radiator, but they require airflow through the fins in order to work efficiently. In any case, moving air increase air transfer coefficient, which means, it helps much better heat transfer.

There are also soldering heat sinks, which protect electronic components during soldering, as soldering is a very hot process. This type of heat sink can be clipped to the lead that is used for soldering which absorbs the heat from it.

Other than heat sinks, for various machines, there are other means of transferring heat, other than using heat sinks. For example, heat can also be achieved through transferring in pipes, liquid or vapor cooling.

Hexapod Robot:

A six legged robot. See *Quadrupeds*, as similar principles apply.

Humanoid Robot:

Please visit the chapter for *Androids / Humanoids*.

Image Recognition / Processing:

See *Object Recognition*.

Impedance:

It is the resistance to alternative current, AC. Impedance has both magnitude and phase, whereas the resistance to DC only has magnitude. The units of impedance is also in ohms, like resistance, but impedance also includes the reactance component, which is the opposition of an electric circuit element to a change in voltage or current.

Inductance:

It is an electrical property of a conductor, where a current change through it, creates an electromotive force, in the conductor itself and in nearby conductors by mutual inductance. For example, if the current flowing through coil of wire changes, an electromotive force is generated to resist this change. Inductance is defined by Faraday's law, where $Emf = - L (\Delta I / \Delta t)$ here Emf is electromotive force in volts, L is the self inductance of coil in henrys, ΔI is the change in current is in amperes and Δt is the elapsed time is in seconds. Here the coil is called the inductor. Also see the term *Electromotive Force*.

Industrial Robot:

Please visit the separate chapter for *Industrial Robots*.

Insulator:

In electronics, an insulator is a component that does not allow the flow of electric current. Insulators are used in electronics in order to separate the individual conductors from each other. Glass, plastics, rubber polymers, teflon are all good insulators. There is no such thing as a perfect insulator and given enough voltage, all materials will conduct electricity, however the materials counted here will require much more voltage than a conducting or semi conducting materials, and therefore they can be considered as insulating for practical purposes.

Integrated Circuit:

It is a body of semiconductor material such as silicon, where the components of an electronic circuit, such as resistors, transistors, capacitors, are completely integrated. Today these are also called chips. Since its invention, integrated circuits revolutionalized the electronics industry, and, they made possible all computers, cell phones, home electronics, robots, and all other electronics that we see around today. Shortly before the invention of IC in late 1950s, it was shown that, the semiconductors were able to perform the functions of vacuum tubes. This knowledge enabled the first *transistors*, and in turn ICs, which then increased in capacity and speed by millions of times, enabling more powerful electronic devices all the time. Also see *Transistor*.

Kalman Filtering:

A Kalman filter is a group of mathematical equations to provide efficient means to estimate the state of a process by minimizing errors. Past, present or future states can be estimated using this method. This mathematical system is very often used in operation of many types of robots, from drones to self balancing wheeled robots, where continuous sensor input must be processed. Kalman filters help robots processors consume less memory by removing noise from the data, because the process is based on weighted average, which gives more weight to data which is likely to be more certain.

Kinematic Diagram:

A kinematic diagram shows how the frame, links and joints and other points of interest of an object or member are connected, disregarding the dimensions of the object, just to show the connections and visualize the system better. As in a free body diagram, the joint conditions must be represented accurately, to allow for only permitted translation movements or rotations, and restrict the ones that are not permitted. For example, a component that sits on wheels must be drawn as a roller joint, where it allows for translation in the roller direction, and rotation at the joint, but does not permit translation in the perpendicular direction. It can also be called as skeleton diagram, kinematic scheme, linkage graph or joint map. Although it may look somewhat similar, this is very different than the *free body diagram*, in which dimensions, forces and moments acting on the object are shown, and the external supports are replaced by the reactions they produce, which is then used to find the reaction forces that act on the body. Also see *Free Body Diagram*.

Kinematics:

It is the study of motion itself, without considering the forces that cause the movement, therefore it only studies the geometry of the movement and can be considered more mathematics than physics. There are mainly two types of kinematics analysis. In forward kinematics analysis, the calculations are simpler, where the final position, velocity, acceleration and angle of a member is calculated, given the position, velocity, acceleration and angles of the other members plus that member and joints which move that member. Here, a good example for a member, would be a robot hand. These calculations are done by matrix analysis in 3 dimension. In the other type, reverse (or inverse) kinematics, the position, angles, velocity and acceleration of the joints must be calculated, in order to reach a certain given robot hand location, angle, velocity and acceleration, as an example, or any system composed of joined parts. The calculations for the reverse kinematics, which is also called kinematic synthesis, are much more complex than the forward analysis. In addition, sometimes the robot arm might have more degrees of freedom than the constraints, which means, more than one combination of joint locations and angles are possible, in order to reach to a given result, which makes calculations even more complicated. Kinematics are also used in other fields of science such as in astrophysics, where the position and velocity and acceleration of planets, stars, comets can be calculated. Also see *Kinematic Diagram*.

Kinesiology:

Kinesiology is the study of body movement mechanics.

KV Rating:

Please see this term under *Drones > Common Terms for Drones* section

Language Perception:

See *Cognitive Computing, Machine Learning*.

Laser Guided Vehicle:

See *Drones and Robotics Vehicles > Automated Guided Vehicle*.

LGV:

Acronym for *Laser Guided Vehicle*. See *Drones and Robotics Vehicles > Automated Guided Vehicle*.

Locomotion:

See *Robot Locomotion*.

Logic Gate:

It is the most basic building block of a digital circuit. It can have a single or more than one input in the form of digital signals, and based on the logic operations it performs, it produces only one output, which is a 1 or 0, or in other words, yes or no, which is the result or "decision" that it produces. It performs these operations by logic operators such as AND, OR, NOT, NAND, NOR, XOR, XNOR. Individual gates can be combined to form more complex circuits or logic gate functions. Logic gates, in the form of transistors, form the basis of integrated circuits. Also see *Digital Input, Integrated Circuit*.

Low Voltage Cutoff:

These are modules to protect batteries from excessive draw. For drones they are a part of ESC, which function when the charge of batteries drop and the voltage is cut to drive motors and supplied to steering servos, in order to safely return the drone to the base.

LVC:

See *Low Voltage Cutoff*.

Machine Learning:

This is a subset of the general field of artificial intelligence, and it means, the ability of a computer to learn from data without being explicitly programmed, by using only the algorithms, in order to parse large amounts of data, learn from it, and make a predictions or decisions, or learn how to perform a specific task or produce some kind of reasoning. Some of the techniques used are reinforcement learning, using decision trees, Bayesian networks and more.

Historically, artificial intelligence as a concept appeared first, in 1950s, then machine learning as a subset of AI in 1980s, and then deep learning, which is a more powerful concept with more uses, as a subset of machine learning, after 2010s.

Deep learning is a subset of machine learning, where it applies a new approach to neural networks and reinforcement learning techniques to solving specific problems. Neural networks and reinforcement learning concepts had been around since very long time, but recently with deep learning, larger networks with more connection points and layers were introduced, which is able to organize data much more efficiently, and combined with the increased computing power, they are able to do much more than the narrow AI. For example, for a neural network, in order to learn a simple task, a value is assigned to each right or wrong behavior. But when the task to be learned gets more and more complex, the assignment of these values become computationally impractical. Deep learning concept solved these by organizing data in a much more effective manner. DeepMind Technologies Inc., owned by Alphabet Inc., the parent company of Google, for instance, which made one of the most famous examples of programs that uses deep reinforcement learning, was able to make its program teach itself to distinguish pictures, by watching millions of images, or learn to play the complex game of Go, by playing it millions of times, and in fact it was able to defeat a human champion in 2016, which made headlines all over the world. This victory was an important milestone, as Go is a game that requires intuition and much more human characteristics than playing chess and it is impossible to program a machine without learning to win this game against human champions. The only way was to program the machine so it does the learning by itself, which it did successfully. Learning is achieved by studying lots of examples, awards for correct behavior, or generalization.

Because of the recent successes, deep learning promises a lot of applications in our daily lives, some of which have already started to appear, such as making recommendations in healthcare or law, autonomous car navigation, applying correct colors to black and white images, predicting a case outcome, recognizing objects that have incomplete or obstructed images, responding to our verbal questions in natural language, playing difficult computer games at superhuman levels in which humans were always able to outsmart computers before, industrial robots learning to perform new tasks without manual programming or human supervision, even help with difficult to solve hypothetical situations such as minimizing energy consumption of a factory by analyzing huge amount of data and making simulations that will produce the best outcomes and many more applications which would probably require many more pages to even write here. With deep learning, it is safe to say that the era of the narrow AI is over. There is still room for improvement in deep learning though. For example, all these successes are possible after practicing or reviewing a huge number of situations or data. In addition, for the situations where more than one goal must be achieved at the same time, there must be new improvements as well.

Machine Vision:

It can be considered as a sub field of *Computer Vision*, please see that term.

Manual Teaching:

To teach a robot intuitively by manually guiding it to do actual desired movements, which usually can be done within minutes, rather than programming with code. This requires an additional add on feature, which is compensated by avoiding the high cost of traditional training with coding, where lots of hours of operator programming is needed. There are certain drawbacks too, as far as the preciseness, which can be corrected by code, after doing the initial teaching.

Mapping:

See *Robotic Mapping*.

Mechatronics:

A combination of electronics, computer, systems and mechanics engineering, computer science. Mechatronics engineers work in many fields including especially robotics and automation.

Medical Robots:

Please see *Medical Robots* chapter.

Microchip:

See *Integrated Circuit*.

Microcontroller:

This is the brain of a robot which receives inputs from sensors or by programming and converts these inputs into commands in order to control anything and take an action, including motors, lights, other electronic devices, or even sending a tweet. In other words, a microcontroller is like a complete computer but on a smaller scale. It can also be defined as a circuit board with a chip that can be programmed to do different things.

It is important to remember that items that require large electrical loads such as motor, solenoids, powerful lights cannot be connected to microcontrollers directly because a microcontroller can only output a small amount of electrical power through its pins, and trying to do so may damage microcontroller. The voltage values of a microcontroller can be seen in specifications. Also, microcontrollers are very suitable for hobby level robotics projects but they cannot perform complex artificial intelligence functions such as intelligent image processing or intelligent navigation or object manipulation, which require very complex algorithms and bigger hardware resources to run.

The core component of a microcontroller is an Internal ROM. The other components are internal RAM, accumulator, timer, counter, interrupt circuits, register, arithmetic and logic Unit (ALU), input output ports. Of these, internal ROM is the read only memory where instructions for the microcontroller are stored. ALU is where the mathematical and logical operations are performed. The mathematical data is stored inside the register. Program counter keeps track of the operations number of programs executed, by counting. Clock circuit generates the pulse for reference for microcontroller operations. Today, microcontrollers are embedded in all kinds of electronic devices.

They also serve as the brains of a robot. For example Arduino, Raspberry Pi, which are two of the most popular platforms used for hobby robotics, are microcontrollers. Also see *Microprocessor, Circuit Board, Embedded Computing, Integrated Circuit.*

A microcontroller board contains:

- Microcontroller
- Input/output pins
- USB connection
- Power jack
- Reset button
- Analog inputs

Following items can be considered when choosing microcontroller board:

- Operating and input voltage
- Variety and number of analog and digital I/O Pins
- Flash memory
- Clock speed
- SRAM
- EEPROM
- DC current
- Dimensions and weight

Microprocessor:

Microprocessors, which are the heart of computers, actually calculate and compute things, are contained within *microcontrollers*, which can be considered heart of *embedded systems*. It is a CPU inside an integrated circuit and this integrated circuit is called a microprocessor. They contain CPU, ALU and a few registers but unlike microcontrollers, do not have RAM, ROM other peripherals. Microprocessors run at much higher speeds than microcontrollers and consume more power.

Military Robots:

Please see *Military, Security, Law Enforcement Robots* chapter.

Modular Robots:

Robots that are made up of similar components, which can form a large number of configurations for different purposes and tasks. Another way of describing modular robots is that they are bigger robots, which are made up of smaller robots. The complexity of a modular robot exponentially increases as the number of modules in it increase. Robot modules are usually in cubic or spherical form but other shapes are also possible. The robots cannot be only made of modules most of the time however. Additions such as cameras, grippers, wheels, mounts are usually necessary, to facilitate functioning of the robot.

Moment:

It is the turning effect of the force acting on an object, which is calculated by multiplying force vector, with the distance to the turning point, such as M = F.d. See the term *Force* for more explanation.

Motion Control:

This is part of automation process, which specifically deals with the moving parts and control of those parts, in order to move a specific load in a controlled fashion, in a hydraulic, pneumatic or electromechanical manner. Components such as motion controllers, actuators, energy amplifiers are used in the process. Actuators vary greatly in speed, power, precision and cost, all suited for different applications. Motion control can be classified into two, as far as how it is done. First is an open loop, where the motion directives are given but the checking of whether the desired results are obtained or not, is not done. In closed loop, the feedback or checking of the result is also performed. For a closed loop system a measurement device is also included, which performs the measurement and sends the feedback to the motion controller, where the controller makes adjustments for any errors. When designing a motion control system, motor sizes, electrical interfaces, servo system integration are designed and combined with motion controller, amplifier and actuator. The criteria or objective of the design include offsets, limits, feed rates, determining machine states.

Motor:

Motors turn electrical energy into mechanical energy, which makes possible movement of robots, along with servos and actuators.

There are different types of motors:

Brushed Motor: This can be considered as the basic type of motor, and the control is as easy as turning a current switch on and off, which enables a certain motor speed. In this type of motor, the permanent magnets are stationary, and the electromagnet is moving. The current is applied to brushes, from the battery, and these brushes, which are stationary, physically touch to commutator, which is moving. So this physical contact of a stationary piece to a moving, constantly turning piece creates friction and heat and causes the brushes to wear out by time. When the commutator takes the current from the brush, it sends this current to the coil of wires which creates a magnetic field, and the permanent magnets make these coils rotate by pushing and pulling them according to their position. The commutator can be considered as a switch, when it turns, the polarity of magnetic field is switched each time a coil passes a magnet. It is this constant switching of magnetic field that creates the turning force. The disadvantages of brushed motors are that they are not very efficient due to friction and heat loss, and brushes and commutators wear out in time. This means, they must be maintained more often. The advantage is that they can be constructed easier than brushless motors and the initial cost of building this motor is lower. Brushed DC motors have two connectors, for positive and negative. To change the direction of rotation, the only thing needs to be done is changing positive and negative from the switch or battery connection.

Brushless Motor: In a brushless motor, as opposed to a brushed motor, electromagnets do not move, but the permanent magnets move, by either being on the inside (inrunner motor) or outside (outrunner motor) of the electromagnets. The movement can be achieved when the magnetic field generated by the electromagnets pushes and pulls the permanent magnets. Therefore at no point for this push and pull process there needs to be friction between pieces which increases the efficiency of these motors, in comparison to a brushed motor. This also means that considerably less heat is generated by these motors. The sequence of providing current and sensing where the permanent magnets are done through ESC or motor controllers and sensors and it is more complicated than constructing a brushed motor. Therefore, the initial effort and cost of building a brushless motor is higher but later it needs less maintenance as there is no wearing brushes due to friction. Most hobby robots and drones today use brushless motors, but very small ones, such as the ones that are less than 6" diameter, may use brushed motors due to simpler controls and lower cost. Brushless DC motors have three connectors which are controlled by ESC. The ESC converts the DC current of battery into three phase AC signals for the motors, which continuously change magnetic field and produce rotation. Therefore, if a brushless motor is connected directly to an unchanging DC power source, the motor will not be able to change magnetic fields and instead of turning it will short circuit and melt.

Inrunner Brushless Motor: The turning piece is the inner shaft and therefore it can turn very fast but produces little torque and therefore needs a gearbox in order to increase the torque delivered. The inner turning shaft is the permanent magnet as this is a brushless motor and the fixed coils are at the outer casing.

Outrunner Brushless motor: This type of motor spins its outer shell, so although slower, it is able to produce much more torque than an inrunner motor. Directly producing a lot of torque in the beginning eliminates the use of gearbox, which reduces weight, noise, complexity, efficiency loss and makes this type of motor suitable for running all types of drone propellers. The outer turning shaft is the permanent magnet as this is a brushless motor.

A typical data sheet for a brushless multirotor motor by manufacturer may look like this:

Throttle (%)	Volts (V)	Propeller (inch)	RPM	Watts (W)	Thrust (g)	Current (A)	Efficiency (g/W)	Operating Temp. (g)
50	22.2	8	1100	40	700	1.7	17.5	39
75	22.2	9	1600	91	1450	3.9	15.9	40
100	22.2	9	1950	141	2050	6.0	14.5	41

One thing to note about this data is that the efficiency (grams / Watts), goes down, as you increase the throttle to 100%.

Below is another example directly from a manufacturer's brochure:

Prop	Volt	Watt	Amp	Thrust (g)	Throttle %	Efficiency (g/W)
HQ 4x4.5 Bullnose	16.5	90.8	5.5	265	50	2.92
Nylon-Glass Fiber	15.8	331.8	21	669	100	2.02
GemFan 5x3	16.5	57.8	3.5	254	50	4.40
Carbon Nylon	16.1	206.1	12.8	660	100	3.20
HQ 5x4.5	16.8	86.7	5.16	342	50	3.95
Nylon-Glass Fiber	16.5	316.1	19.16	802	100	2.54
GemFan 5x4.5 Bullnose	16.6	114.5	6.9	376	50	3.28
Nylon-Glass Fiber	16.4	373.9	22.8	866	100	2.32
HQ 5x4x3	16.6	112.9	6.8	426	50	3.77
Nylon	16.4	406.7	24.8	962	100	2.37

Brushless Motor. The chart is a typical data sheet that shows different efficiency values for various propeller sizes, voltages and throttle percentages. Photo Credits: GetFPV - www.getfpv.com

Stepper Motor: A stepper motor is a type of DC motor that rotates not continuously but in controlled steps, which divides a full rotation into the desired number of steps. They consist of a magnetic rotor shaft which rotates, and surrounding stationary electromagnets which causes this rotation, in desired steps. When the electromagnets are energized in steps, the rotor, which is a magnet, responds to changing magnetic field and rotates in controlled steps. Using stepper motors allow precise positioning of motion, as opposed to constantly rotating motors. As the angle between each step decrease, the resolution of the motor increase. For instance, a stepper motor which has 4 degrees between each step needs 360/4 = 90 electromagnets around the rotor, to create each step. There is no need for a feedback mechanism for accurate control, as long as there are required number of steps, and activation of magnets are performed accurately, but a feedback loop can also be useful, in cases such as returning the rotor to original position, in case of a rotation caused by an external force such as by hand. To control stepper motors, two types of signals are required, such as the direction and pulse frequency. The precise control of rotation is a very useful feature for motion control and therefore stepper motors are very frequently used in countless types of robots and machines.

In addition to the general categories of motors that we mentioned here, there are much more types of motors, with very different characteristics and uses, but that is out of our scope. Also see *Servo, Motor Controller & Driver*.

Motor Controller & Driver:

A motor controller, regulates acceleration and speed of the motor, by either supplying digital output for a stepper motor, or, low voltage analogue output for a servo motor. This output current may be high enough to drive a motor directly, plus the inductive reaction from the motor can damage output circuit. So, a motor driver is included, which can amplify the voltage and current to drive a motor. This enables a motor controller to control a variety of motors, with the help of different motor drivers.

Therefore, a motor drive controls the electric current that is sent to a motor. It gives current to the motor in different amounts and frequencies to control motor's speed and torque. A standard inverter drive controls speed and torque only, whereas a servo drive controls speed and torque, as well as positioning machine components for applications that need complex motion control.

Motor controllers and drivers accept different kinds of signals. Motor drivers use low level control signals, such as PWM and direction, while controllers have on board microcontroller to use low level signals plus accepting high level signals from the user, such as TTL serial signals or USB communication signals. Motor controllers can have features like current sensing, acceleration limiting, limit switch support and feedback based control.

A brushed DC motor driver is the simplest type, as all they do is to amplify signals for a DC motor, such as PWM and direction, so the low level, resource consuming signal generation that will be supplied to the drive is taken care of by the controller. A stepper motor drive can control stepper motors, which are more complicated to control as they require adjustable current control, direction inputs and multiple step resolutions.

Also see *ESC*, which is more often associated with RC controls, and *Servo Controller & Driver*.

Nanobot (Nanorobot):

These are the nanoscale robots, that are made from atoms or molecules. This is an emerging field and there are not any nanobots yet that are in practical or commercial use but there is ongoing research. When built, nanobots are expected to find countless applications in many industries, especially in medicine. Nanobots which are able to reproduce can also pose a significant threat if uncontrolled.

Natural Language Processing (NLP):

Natural language processing extracts patterns to better understand the structure, sentiment, polarity, style of writing, contextual information, and much more. It is the way of communicating with AI not by writing code but through using plain human language such as English, where the AI processes it mathematically and algorithmically into the code that it can understand, and gives back the output again in natural language, or takes appropriate action by "understanding" the question or command.

Although this is a stand-alone field, many natural language processing problems use machine learning / deep learning principles, and therefore there's considerable overlap between NLP, machine learning and deep learning. For example, by using deep learning principles, the initial observations from the pre-processing of a natural language document term matrix/corpus are strengthened, and this enables better understanding of new bodies of text based on classification, similarities, distance measurements and more. Principles of NLP can be used for things such as predicting the behavior of masses, summarizing articles, search engine recommendations. Also see *Cognitive Computing, Machine Learning.*

Navigation:

See *Autonomous Navigation.*

Negative Feedback:

It is a term used often in motion control, where the sensor sends feedback to the motor that it is yet to make movements in order to reach to a desired position. As the motor moves toward the desired position, the amount of negative feedback decrease, telling the motor that it is moving more and more in the right direction so the motor turns slower as it reaches more toward the desired result and stops when the negative feedback ceases.

Neural Network:

It is the artificial modeling of brain neurons to solve problems in the same way a biological brain does. Many artificial neurons are connected to each other as a big network, and, when working on learning a specific task, some connections are reinforced if there is a positive outcome, and some are inhibited if negative. This enables self learning of a specific task. It is now known that neural networks are not even close to being sufficient to model the brain and we need to make more improvements to reach a human level AI. Greater complexity is needed and further studies are performed such as linking layers of neurons, in addition to just neurons, or, sending different types of signals rather than just on or off signals, or dynamic networks which can make new or destroy existing connections. Therefore, although regular single layer neural networks cannot adequately model the brain, they serve as a base for other approaches. Currently, even the most advanced neural networks still have millions of times less connections than the human brain, but this gap is decreasing.

The concept of deep neural networks were introduced more recently, where a multi layered neural network is introduced, instead of the usual single layer. This enables more complicated tasks to be performed, by modeling complex data with fewer units than possible with a single layered network. Deep neural networks are effective in learning hierarchical feature representation. This means, they first look at the fine details of the whole picture, then combine them into a higher hierarchy, then higher, until the whole picture can be understood, in a much more effective way than single layered networks. This characteristic is very effective means of recognizing things, because to understand objects, computers must first look at the finer details. DNN forms the basis of deep learning, which is a very promising concept and can be considered the most capable sub field of AI as of today. Also see *Machine Learning, Cognitive Computing,* and the separate section of *Artificial Intelligence.*

Neuroprosthetics:

A neuroprosthetic device replaces or supplements inputs and outputs of nervous system. These devices can be used in conjunction with robotic limbs in human body.

NLP:

See *Natural Language Processing*.

Object Recognition:

It is the process of identifying objects from the photos or videos. Algorithms aim to enable the machines to self learn to identify objects, using techniques such as recognizing patterns, edges, gradients, feature detection, feature extraction, searching model databases, segmentation of images. Objects can appear from countless of different sizes and angles and lighting conditions and they can also be partially covered. These are big challenges in object recognition. Identifying objects can include a lot of different abilities such as identifying traffic signs or objects, detecting faces, or even facial features. Object recognition is a sub-field of *computer vision*. Also see *Object Tracking*, *Computer Vision*.

Object Tracking:

For mapping of the environment and autonomous navigation, the objects entering into a camera vision must not only be identified, but also be tracked as they move. The challenges to track objects include changing lighting and angles of moving objects, obstructions, camera movement, loss of information when projecting 3D images onto 2D, real time processing requirements, object shapes that deviate from normal, objects that display complex motion, and more. The process can be simplified by making assumptions and putting limitations to path of the objects and their appearance. There are a large number of tracking methods, for different situations. A lot of questions such as which image features should be used, how should the motion and shape of the object be modeled must be answered, and different needs and environments and devices and hardware also come into play. Also see *Object Recognition*, *Computer Vision*, *SLAM*, *Autonomous Navigation*, *Robotic Mapping*.

Obstacle Avoidance:

For mobile robots, avoiding collusion and moving around obstacles or other moving objects is one of the primary tasks to be fulfilled. Obstacle avoidance techniques are used on almost all robots ranging from hobby robots smaller than a human hand, to autonomous cars that carry humans. In the most basic sense, avoiding an obstacle consists of identifying an obstacle with a sensor such as infrared or ultrasonic, and, if the identified obstacle is closer than a certain distance, adjusting the path of the robot in order to avoid collusion with the obstacle. Also see *Autonomous Navigation*.

Ohm's Law:

Ohm's law states that current through a conductor is directly proportional to the voltage. This proportion is the resistance, which is stated by R = V / I, where R is the resistance in ohms, I is the current in ampere and V is the voltage in volts. Also See *Current*.

Ohmmeter:

A device which measures electrical resistance.

Oscillator:

It is an electronic circuit that produces oscillating electric currents or voltages in the form of sine or square waves.

Oscilloscope:

It is a device that measures varying voltages.

Parallel Robot:

It is a type of industrial robot, where several arms work in parallel. Also see *Articulated Robot, Cartesian Robot, Scara Robot*, for other types of industrial robots. Also visit *Industrial Robots* section.

Passive Dynamics:

It is a legged robot walking mechanism that can walk downwards by itself, just by gravity, without drawing energy from a battery and without the need of an actuator or motor.

Pattern Recognition:

This is similar to machine learning, both use data mining techniques, but pattern recognition is more based on statistics, while machine learning primarily uses algorithms to make meaning out of data. Also see *Machine Learning*.

Payload:

It is the maximum load an industrial robot can carry, to perform its tasks, without losing its precision for the desired path of movement. Factors such as capacity of the motors or actuators, acceleration requirements, length of robot arm, friction can affect this value. For an aerial drone, it means the load carrying capacity of the drone.

Physics Engine:

A software that creates a virtual physical world, which aims to simulate the real life physics in virtual reality. Robotics simulators use and integrate with physics engines in order to define the environment that the virtual robot will operate. Also see *Robotics Simulators*.

Programmable Logic Controller:

This is a computer that is specifically designed for use in manufacturing environments to control robots. It uses a programmable memory to store and implement specific instructions, for logic, sequence, timing and more. As in any robotics system, the inputs to the computer comes from the switches and sensors, processed by the software and the outputs are sent to motors or in terms of signals to different locations.

PWM:

Acronym for Pulse With Modulation. It is a technique for controlling analog circuits, to encode message into a pulsing signal, using digital output from microprocessors. It has a great number of applications. For example, an ESC, which is used to control motors for RC robots and drones, works by receiving PWM signals, to start operation. Using PWM signals brings down the system costs and power consumption significantly.

Quadruped Robot:

It means a robot with four legs. Legged robots are more difficult to build and consume more power than wheeled robots, but they have certain advantages such as ability to advance on irregular surfaces, or climb steps. The walking gait of quadruped robots can be analyzed using inverse kinematics concept. Number of servos (it differs between a quadruped or hexapod etc..), values of servo minimum and maximum points, the pose at which the robot is neutral, and the correct direction of movements (sign testing) must be defined in the analysis program, which then interacts with the gait engine. Also see *Kinematics*.

Reach:

It is the distance from the base of the robot arm, to the edge of its wrist. For most industrial and collaborative robots that are in the market now, this value usually varies anywhere between 500 mm to 2000 mm. Also see *Industrial Robots*.

Reactance:

It is the opposition of an electric circuit element to a change in voltage or current, which arises as a result of that element's capacitance (for resisting voltage change) or inductance (for resisting current change). Also see *Current, Electric Circuit*.

Reinforcement Learning:

It is a system that learns by favoring the behaviors that led to positive outcomes, when learning a task, all by itself. This term relates very closely with *Deep Learning, Neural Networks*.

Relay:

A relay is a switch that is operated electrically, with the use of electromagnets. It is made of electromagnet, spring, armature which the electromagnet attracts, and electrical contact points. A relay logic is a schematic representation of relays that are connected to each other in any combination desired.

Remotely Operated Robot / Vehicle:

Not all robots are autonomous. For tasks that require complicated thinking or navigation, a robot can be remotely controlled, where the tasks or navigation can be performed from far, especially for the locations that humans cannot reach or would rather not go into, such as operating in a hazardous environment. A drone is any robotic aerial, land or sea mobile robot which is remote or auto controlled. This is contrary to the general belief that a drone means only a flying robot.

Repeatability:

An important criteria for industrial robots, repeatability means how well the robot will return to a previous position that is recorded in controller memory. For most industrial and collaborative robots that are in the market now, this value usually varies anywhere between +/- 0.02 mm to +/- 0.2 mm. The difference of repeatability from accuracy is that, accuracy measures how well the robot reaches to an external commanded position, where repeatability is how well the robot returns to a position and orientation in its own controller memory, which is a memory that was created when the robot was there before. This is a very important criteria for an industrial robot as it defines how well the robot can repeat the same task. Also see the term *Accuracy*, and the figure presented there.

Resistance:

The measure resistance of a conductor to passing of electrical current though it. Dividing the voltage by resistance, gives the current value, as in the equation I = V / R. Also see *Current*.

Resolution:

It is the smallest incremental move that the robot can physically make. It is one of the factors that affect the accuracy of the robot, which is an important criteria for industrial robots.

Robot:

The word robot comes from the Slavic word "rabota" which means "work". A robot is a machine that can perform a task or a series of tasks automatically or with human guidance. Robots manipulate objects by moving, grabbing, they can move from one location to another, they can perform dull, repetitive or dangerous tasks with greater accuracy, faster and cheaper than humans such as in industrial applications, they can clean floors, manufacture cars, pick fruits, assist elderly, play ping pong, perform surgeries, move around buildings as security guards, provide surveillance, organize warehouses, deliver goods, diffuse bombs, fly, drive or swim autonomously, explore surface of Mars, help in children's education, clean up hazardous waste, explore mines, and do countless more tasks. The tasks they perform can be as simple as following a line, or as complex as performing sensitive operations in a nuclear plant.

Again, a robot does not need to act automatically to be classified as a robot, and machines that can only act with human guidance, such as remote controlled machines also can be classified as robots.

The robots that act automatically can be further classified into two groups, the ones that can only follow a specific set of orders that were manually coded beforehand, such as industrial robots on a welding line, or, the ones that can do things that were not coded before and can act and learn on their own depending on the new circumstances, by using their artificial intelligence and learning algorithms. More and more robots are starting to fall into the latter category as there are improvements in hardware and artificial intelligence. Therefore the science of robotics closely rely on software and artificial intelligence, as well as kinematics, electronics, computer engineering, mechanics, and in some cases even biology.

Robots can also be classified into many groups as far as what they do and how they do it. Some robots are as small as grain of rice, while others can be much bigger than a human. Many robots act as single entities, while there are also robots that act as a team, in swarms. The robots have long been in industrial plants, performing operations that needs speed and precision with prior programming, and due to improvements in technology, they are now at a stage of entering our daily lives, where they can act in previously undefined environments, and we are starting to see more of them around us.

Soon robots will be everywhere in our daily lives, reshaping our civilization, and we will own robots as we own cell phones today, with different models, running apps for different tasks, probably with few standard widely used operating systems and several big companies manufacturing the majority of them. Robots will not cause unemployment overall, they will simply cause humans to work on other, higher quality, more information based jobs. All of us owning robots in the future, to help us around with our daily tasks that we used to perform, implies that these robots probably will have to be close to a human form, such as androids / humanoids. This book introduces 8 different categories of robots in separate chapters.

Robot Apps:

Similar to phone apps, when personal robots become mainstream, robot apps are expected to be used widely, and be downloaded to robots to perform different tasks and give them different abilities, just like the phone apps of today. They have already started to emerge, for a limited number of robots and tasks.

Robot Cell:

Robot cell is a work location dedicated to a robot or robots to work. It includes a complete assembly of robot, all of its process equipment, safety equipment, all peripherals such as vision systems, welding positioners, robotics software, grippers for a complete self contained and ready-to-go environment for manufacturing, that is pre-engineered so that very little or no custom engineering is required. The layout is carefully designed to optimize the production output based on the demands. The location of robot, products, entry and exit point to the cell, location of electronic equipments and other electronic or mechanical components are coordinated with each other when making the layout. A single cell is usually made for a single robot but cells that include more than one robot configurations are also possible.

Robots in work cells can do different operations, such as welding, packaging, pick and place, palletizing, and anything that industrial robots do. When a work cell solution is introduced to an existing manufacturing facility, the transition needs to be smooth in order not to interrupt production, or with minimal impact. A cell is surrounded by fences to keep people from entering for safety reasons when the cell is operating. Another safety measure is a laser scanner, where the work cell is constantly scanned by laser to make sure that the area still fits the description of a predefined area, and when it is not, it means an intruder is in the cell, that causes the operation to stop immediately.

Robot Development Platform:

Robot platforms are used to build robots on them, serving as a base. These robots can range from simple educational robots for kids, to advanced robotic systems for complicated missions. For hobby level platforms, many of these boards can work with common and open source microcontontroller boards such as Arduino or Rasberry Pi. The board of the development platform have interfaces for these boards, various sensors, motors, batteries, switches, tracks or wheels, and therefore they serve as a base for developing many different applications.

Robot Hand:

Hands are the most important part of a human body to manipulate objects. To make robots that can perform tasks and manipulate objects in a similar way to humans, it is necessary that robots have hands with equivalent abilities. Robot hands are a sub group of *end effectors* which include other types of manipulators especially for industrial robots, such as grippers or vacuums. For medical robotics, robotic hands are also used often. There are also robot hands made for amputees. On the right, you can see an example of a robotic hand.

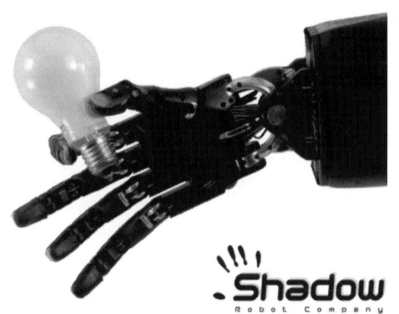

This is a dexterous robotic hand, which is built with an anthropomorphic approach. With 20 actuated degrees of freedom, absolute position and force sensors, and ultra sensitive touch sensors on the fingertips, the hand provides capabilities for problems that require very close approximation of the human hand. It uses industry standard interfaces and can be used as a tele-operation tool or mounted on a range of robot arms as part of a robot system. Photo Credit: Shadow Robot Company Ltd. - www.shadowrobot.com

Robot Kit:

Directed mostly for beginners in robotics or kids, robot kits has various components that can be arranged in lot of different configurations in order to make different robots. They are sold in complete packages that include motors, mechanical parts, sensors, coding interface, microcontrollers. Many robot kits can be considered in toys category but there are many kits that are directed towards hobbyists also.

Robot Locomotion:

How a robot moves around is called robot locomotion. This can be achieved by different systems, such as legs, tracks, wheels or a combination of these.

For legged locomotion, the complexity of the robot increases exponentially, with increasing number of legs. If two legs are used, such as in a humanoid robot, it is called a bipedal, if four are used, it is called a quadruped, if six legs it is called hexapod. These three are the most commonly used systems for a legged robot. The biggest challenge for a legged robot is achieving stability while moving legs and the robot. Legged robots are used when the robot needs to perform climbing, walking on irregular surfaces, or even jumping. For legged robots, efficiency of locomotion is an important factor, which is directly proportional to how much weight the robot is able to transport at what speed, and inversely proportional to the amount of power it uses. Also see *Quadruped Robots*.

Wheels are more power efficient, cheaper and far easier to construct than a legged robots, but may experience difficulty or sometimes even find it impossible to move on irregular or rough surfaces. Apart from the using 2 or 4 wheels as in almost all cases, it is also possible to have different configurations, such as an omniwheel drive, which can provide very precise movement in all directions but difficult to control. It is also possible to combine the advantages of wheels, with moving shafts or even with legs, such as in planet exploring robots which has many wheels that can also move up and down independent of each other, to move on very irregular surfaces, or a robot with legs and wheels which can move fast and the same time jump, climb or move on very irregular surfaces, such as the Handle robot made by Boston Dynamics company in the US, which can be viewed at: https://www.bostondynamics.com/handle. These combinations, although much more difficult to build and require very advanced algorithms, can produce superior results in comparison to a system with only legs, wheels or tracks alone.

Tracked motion offers a great deal of stability while moving. As far as moving on irregular or rough surfaces, it lies somewhere between the legged and wheeled robots. Sometimes out of ordinary configurations of tracks are used, in contrast to the usual two tracks, in order to achieve greater flexibility and moving ability, but of course this comes with a cost of power consumption as well as greater control difficulties. Robots used in defense and law enforcement sectors, use tracked systems often.

Robot Navigation:

See *Autonomous Navigation*.

Robot Peripherals:

These are add-ons to an industrial robot system, in order to improve or add abilities to a robot. Examples include vision systems, welding positioners, robotics software, grippers.

Robot Platform:

See *Robot Development Platform*.

Robotic Arm:

Here we mean industrial robot arm, and not the bionic arms made for people. For that subject, please see the term bionics. Industrial robot arms are used for a great variety of tasks in industry and even in consumer robotics applications, in order to perform tasks with speed, accuracy, precision and in a cheaper way. They can have grippers, tongs, pinchers, suction cups, to manipulate objects or welding attachments to do welding, cameras and sensors can also be attached depending on desired use. All of the joints that lead to the end, in other words, all joints on the arm is called kinematic chain. The part that is between the arm and the environment, in other words, the "hand" of the robot is called the "end-effector". Robot arms operate by turning around their base, at their elbow, at their wrist and also the gripper locations. The motors of a robot arm get progressively smaller from the base to the end effector, as the motor at the base needs to handle the greatest moment (moment = force x distance, therefore the motor at the base carries the greatest load because it carries the greatest distance).

A robotic arm is primarily specified by its *DOF (degrees of freedom)*, its weight lifting capacity, its range of motion such as vertical and horizontal reach and rotation limits if any. The weight lifting capacity is given with respect to the position where the arm is open the most, as the worst case for the statical moment magnitude it can support. The total DOF of a robot arm is the sum of DOF of all joints. Apart from industrial use, robot arms are also common in robot kits, for any level of domestic or hobby or even toy robotics projects. Some robot arms can also be attached to robot platforms. Like operating many other type of robots, ROS - Robot Operating System, which is an open source and worldwide used software for robotics, is frequently used in operation of robotic arms. For bionic arms that are used as prosthetics for medical purposes, see the term *Bionic*. Also see *DOF, Kinematics, Industrial Robots, Robot Cell, Robotic Manipulation*.

Robotic Manipulation:

Manipulating objects is the most important and common reason to use robots. A robot arm has an *end effector* attached to it, to manipulate objects. A biologically inspired multifingered robot hand gives the best results for dexterity of the robot hand. Robot manipulation is not only important in industrial applications but medical field as well. Principles of rigid body motion with 3 dimensional vector analysis for both translational and rotational movements are applied in studying robot manipulation and robot kinematics and requires highly complicated mathematics. Today, missing limbs is a problem that many people face, and improvement of robotic manipulation technologies will greatly improve lives of these people. Also see *Robotic Arm, Kinematics*.

Robotic Mapping:

Most of the time, unless we are talking about robots that perform tasks in their fixed location, robots move around, either by remote control or autonomously. During moving, especially in the case of autonomous navigation, robots need to localize themselves in the environment they move. This can be an indoor environment, for a domestic / service robot, or an outside environment, such as a car.

For a robot, there are two methods to follow, when localizing itself. The first source is internal, which is the dead reckoning principle, but this principal is not very reliable and only must be utilized as a single source only if outside data from sensors is not available. In dead reckoning, the movement is tracked only by internally, such as by measuring how much the wheels of the robot has turned, such as an odometer measurement, or for how long and at what airspeed a drone has travelled, without considering any outside check. Of course, even the slightest errors will accumulate over time using this method, as the robot has no way of checking it with respect to real world and therefore, this method is only used for a limited amount of time such as when both the GPS and sensors are disabled for some reason. An example might be when a drone enters a tunnel and the GPS data is not available.

The second source of information is external, which is the use of various types of sensors and even GPS data for outdoor applications. This is the main method that is used by the robots for localizing themselves and drawing maps of surroundings. Sensors such as ultrasonic, LIDAR, infrared, GPS are used and the data is fed into a mapping software. Here is an example for basic mapping program:

- The sensors scan the entire 360 degrees of the surroundings, by rotating
- The distances measured are converted into polar coordinates
- The coordinates are assigned probability values of occupied or non occupied by the software.
- Repeated measurement refine the probabilities and a map is created for the area, for occupied and non occupied points
- Using this data the robot can navigate itself, by moving around, while updating map

The concept that the robot navigating itself while constantly updating the map is known as SLAM, which is the acronym for Simultaneous Localization and Mapping. Please see the term *SLAM*, where this is explained in more detail. This sequence as described above was only a simple example from a very basic mapping system. Robotic mapping is one of the most intensely studied areas of robotics now, and a great deal of advanced software is developed for various purposes, from locating a vacuum cleaner, to navigating the autonomous cars. Also see the terms *Autonomous Navigation*, *Digitizing* and sections *Drones & Robotic Vehicles > Unmanned Ground Vehicles > Autonomous Cars*, and, *Drones & Robotic Vehicles > UAV > Common Terms in relation to Drones*.

Robotic Process Automation (RPA):

This is actually a term in business, which does not involve physical robots, which is used for automating certain tasks of people to be more efficient with the help of computer software. It aims to create a virtual workforce, which is designed based on certain business related office tasks.

Robotic Surgery:

Today robots are used in many types of surgeries as they can reinforce capabilities of doctors with their precision, also enabling less invasive surgeries. The surgeon controls robotic arms remotely and his movements are translated into more precise movements. More precise and less invasive surgery means less blood loss, smaller incision and quicker healing time. It has the disadvantages of cost and usually longer operation times. Also see the term *haptics and* the chapter called *Medical Robots*.

Robotics:

It is the study of robots, and the industry that manufactures, sells and uses them. It is also a new area of study in universities, made up of mechanical, electrical and computer engineering, and getting more popular every year, as robots increasingly enter in our lives. Also see *Robot*.

Robotics Simulator:

To design robots can be much more cost effective and faster, if simulators are used. Robotics simulators are currently used more for industrial robots, which operate in predefined environments, and mostly with preprogrammed motions. This is much easier to simulate than sensor based robots in everyday environments, which operate based on instantaneous sensor input and in real physical world, which is not a predefined environment. To operate a robot simulator will also need a physics engine, which will provide the virtual physical world, that the robot will operate.

There are also many recently developed applications which are designed to help average user to design robots. One such application is Carnegie Mellon University's interactive design tool, which helps even an inexperienced user to put together pieces in order to design a prototype of a robot.

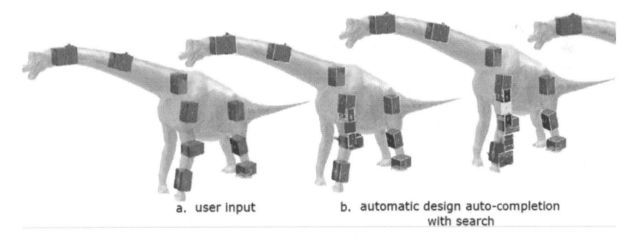

a. user input   b. automatic design auto-completion
                          with search

An automated design tool developed at Carnegie Mellon University allows users to start with a mesh of the robot they want to build and specify the locations of motors using a drag-and-drop interface. The tool then suggests pieces that would connect the motors and allows users to add body plates and other pieces that complete the robot. See next figure for continuation. Photo & Caption Credit: Carnegie Mellon University - www.cmu.edu

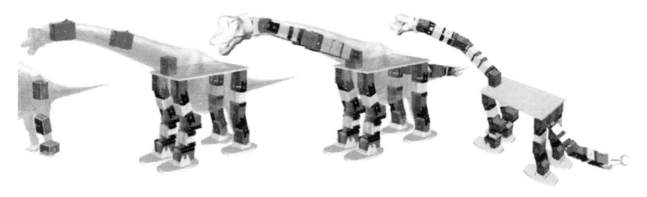

letion   c. adding body plates   d. final result
                    and end-effectors

Continuation of previous figure. Image Source: Carnegie Mellon University - www.cmu.edu

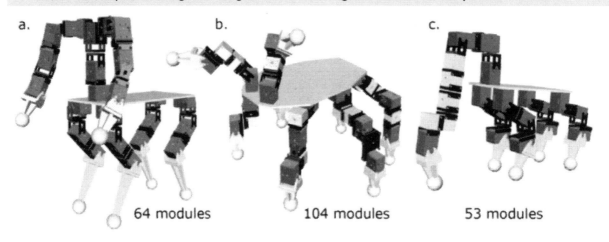

a.   b.   c.

64 modules   104 modules   53 modules

Various designs with different number of modules. Image Source: Carnegie Mellon University www.cmu.edu

Robo-calligrapher

Puppy

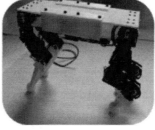

To demonstrate their automated robot design tool, Carnegie Mellon University researchers designed a wheeled robot that can write with a pen and a four-legged "puppy" robot and then followed the designs to build actual robots with 3D-printed parts and off-the-shelf components. Photo & Caption Credit: Carnegie Mellon University. www.cmu.edu

Robotics Software / Programs:

Please see the separate chapter called *Software*.

ROV:

Please see *Remotely Operated Robot / Vehicle*.

RPA:

Acronym for *Robotic Process Automation*.

SCARA Robot:

A type of industrial robot, SCARA is the acronym for Selective Compliance Assembly Robot Arm. It is fixed in z axis, but can move in x and y axes, which is useful when working on a single plane, without moving the robot arm vertically. Also see *Articulated Robot, Cartesian Robot, Parallel Robot,* for other types of industrial robots and visit *Industrial Robots* chapter.

Self Guided Vehicle:

See *Drones and Robotics Vehicles* > *Automated Guided Vehicle*.

Self Learning:

When used for AI, this means that the AI program (or the robot that operates with the AI program) starts with a routine but based on the most successful outcomes, it adjusts its actions to improve itself. For more, see *Machine Learning*.

Sensor:

For any term, concept or type related to sensors, please see the chapter called *Sensors*.

Servo:

A servo is a motor that is also able to turn its shaft within a controlled rotational range, such as 180 degrees. Unlike a DC motor which turns continuously, it has a control circuit, position sensor (a potentiometer performs this task) and gear set that regulates its motion within desired ranges of motion. Therefore servos are very useful in robotics, such as moving robot arms, legs and anything that requires a controlled and accurately positioned range of motion. The position of the servo is attained by sending the servo a coded signal in order to bring it to a certain position and as long as the same signal is present, the servo shaft will maintain that position, and it will start moving again when the coded signal starts to change. A typical control pulse as an example, can be every 10 milliseconds, to tell servo what to do, even if it tells to stay in the same position. Depending on the torque rating of the servo, it will move to the commanded position, even if a force is applied against it and try to hold it. Similarly, it will resist to move away from that position, as much as its torque rating allows, if the coded signal is telling the servo to stay in the same position.

With respect to stepper motors, servos offer higher controlled and more flexible torque, higher speed and acceleration, but only within a limited angle, while stepper motors always operate at their full torque, and have higher holding torque. Servos need a feedback loop mechanism to control positions, unlike stepper motors. Servos, which are usually more expensive than stepper motors, are better suited for situations with dynamic load changes, such as controlling robot arms and legs. Servos have far less poles than stepper motors.

Servo Controller & Drive:

Servo controller circuit board that can control multiple servos by controlling their speeds and ranges. Servo drives are used for precise speed control, for high resolution RM changes. They basically serve the same purpose for *motor controllers and drives*, as described under that term, which differences specifically to control servos. Below is an example.

Miniature servo & stepper motor drive that has integrated multi axis motion controller.
Photo Credit: Granite Devices Inc.
www.granitedevices.com

Servo drive motherboard for embedded robotics such as various table sized robots, pick and place robotics, CNC or 3d printers. Photo Credit: Granite Devices Inc.
www.granitedevices.com

Servo Horn:

It is the moving, turning part of the servo, it can be metal or aluminum, it can be of one, double or four arms.

Servo Mount:

A piece that is used to attach servo to the frame of a robot.

SGV:

Acronym for *Self Guided Vehicle*.

SLAM:

SLAM, which is the acronym for Simultaneous Localization and Mapping is the robot navigating itself while constantly updating the map. This concept is actually a little bit of a chicken and egg problem, because the robot itself moves when trying to draw the map, but, for the robot to accurately localize itself, it needs to know the environment more accurately, which creates a paradox. There are probabilistic methods to tackle these problems such as Monte Carlo method or the Kalman filtering, which attempt to approximate the location of the robot with an acceptable error margin. When making algorithms for mapping, one of the main driving factors is the type, power and limitation of the sensors. Different sensor types with different advantages, measuring ranges, speeds give rise to different software. During SLAM, an observation may include a laser scan, a camera image, ultrasounds or infrared scan or more than one type of sensors to complement each other. The desired outcome is the map of the environment and the path that the robot must travel, which it calculates autonomously.

During SLAM, the robot is located with respect to landmarks in the environment at different times, such as the trees around the robot. The location of landmarks are known points and therefore robots use this continuously to accurately locate itself. This is similar concept of a car which uses a simple GPS system locates its position relative to satellites, or a construction surveyor determine a location or an elevation based on a measurement with respect to a known, fixed, benchmark point, such as the face of a monument or a corner point of an existing building. Of course at least two landmarks are needed for accurate positioning, otherwise, a whole circle is possible around a single landmark, but only when this is compared with the second landmark two possible circles intersect and becomes a known point to locate the robot. The robot then continues to refine the map as it passes through same or nearby areas. Also see *Robotic Mapping*.

Soft Robots:

It is possible to make a clear distinction of the term soft robots or soft robotics into two distinct categories. The first category, can be considered as robots which include soft materials for certain applications. These are already used in real life applications such as the *gripper* shown below.

Gentle handling through the use of soft grippers are necessary when dealing with delicate materials, such as bagels as shown here.
Photo Credit: Soft Robotics Inc.
www.softroboticsinc.com

The second main category of soft robots can be described as robots that have the entire body made from soft materials, such as silicon based materials, flexible actuators, including air muscles or polymers that can be activated with electricity, so these robots are entirely soft objects. This category is still in research phase and has no real applications yet. As an example, In 2016, scientists from Harvard University, created a robot called "Octobot", mimicking the shape of an octopus, which was a robot that was completely made out of soft parts. According to HARVARDgazette website (http://news.harvard.edu/gazette/story/2016/08/the-first-autonomous-entirely-soft-robot/), the robot weighs just 6 grams and is made of silicone, has pneumatic 3D printed chambers inside. The robot propels itself with hydrogen peroxide, which interacts with flecks of platinum inside the robot and produces gas from a chemical reaction, which then flexes or contracts the robots pneumatic 3D printed chambers that serve as muscles. The next step is the addition of sensors so that it can have more abilities such as navigation, and make the fuel last longer. Also see *Biobots, Artificial Muscles*.

Software:

Please visit the separate section called *Software*.

Solderless Breadboard:

Please see *Breadboard*.

Speech Recognition:

This is often confused by voice recognition. In speech recognition, the computer understands *what* is spoken, whereas in voice recognition the question is *who* is speaking. Here the system does not require training, as who is speaking is not important. Speech recognition is much harder to develop than voice recognition systems.

Static Balance Control:

This study must be done for legged robots, such as bipedal robots, to keep its posture in balance and not to fall, while at rest. To keep balance while at rest may be seen as an easy and static task, but actually even for a human, it is a dynamic process where our brain constantly keeps our balance, even if we are only standing. For a humanoid robot, to keep dynamic balance, there must be sufficient power in actuators or motors supporting the balance, sensing information regarding posture, and acting on it with correct algorithms and motion control in order to keep balance is needed. Also see *Dynamic Balance Control, Zero Moment Point*.

Swarm Robotics:

Mimicking the behavior of natural swarms such as bees or ants, the study of swarm robotics aims to produce useful behavior from collaborative efforts of many robots. The swarm can act as a whole in order to accomplish tasks that individual members cannot do. Individual members of a swarm are usually simple but as a swarm they can exhibit very complex behaviors. Note that the shape and form of individual robots in a swarm do not need to be like the ones shown here, and they can be of any shape or size, for any task on land, sea, underwater, air, or indoors, although the robots that are shaped similar to the ones shown as here are arguably the most common. Controlling swarms through a central computer is usually the most straightforward approach but it is highly calculation intensive, and as the swarm gets bigger, this way of control gets increasingly difficult. Instead, writing intelligent algorithms that can improvise and adapt to new situations are more preferable, although programming of such swarms are more difficult.

A swarm example.
Photos By: K-Team Corporation
www.k-team.com

Some example projects or uses of swarm robotics include:

- Cleaning up or monitoring oceans
- Warehouse robots
- Underwater search
- Complimentary military force to swarm enemy targets (might be a major force in the future)
- Making entertaining shows in the air with drones
- Autonomous cars communicating with each other in order to maximize efficiency of road use
- Agricultural applications
- Painting, repairs, cleaning of large areas

Synthetic Skin:

See *Electronic Skin*.

Telepresence Robot:

Please see *Domestic Robots > Telepresence Robots* section.

Telerobot:

A robot that is controlled remotely from a distance. By definition drones are also controlled form distance, but the variety of robots meant for telerobots are not just drones, they can be any kind of robots. Also see *Drone, Telepresence Robot.*

Thermistor:

It is a resistor that changes its resistivity with changes in temperature.

Torque:

It is the turning effect of a force, multiplied by distance just like moment. It is an important criteria of motors and servos, which is always given in the specification sheets by the manufacturers, which indicates in what quantity a turning effect that motor can cause or tend to cause or hold. Rotation may or may not happen when there is a turning effect and it doesn't affect the value of torque the motor can produce. If the produced torque can overcome the resisting reaction, there will be rotation. If not, the motor will still force the body to rotate, but the body that is acted upon will not rotate. Please also see *Moment* and *Force*.

Transducer:

This term is explained under *Sensors* chapter.

Transistor:

They are the true basic building blocks of modern electronics today, found in microprocessors, which are the brains of integrated circuits of any electronics device. Transistors are made from semiconductor materials like silicon and three terminals. Their number in an electronic device may vary from a few to trillions. They can close or open a circuit, therefore act as a logic gate, or linearly control the resistance amount. They also have the ability to amplify electronic signals. So unlike a resistor, which only enforces a linear relationship between voltage and current, transistors are non linear devices with different modes of operation. The first mode is saturation, where the current freely flows and transistor acts like a short circuit. The second mode is cut-off, where transistor acts like an open circuit and therefore allows no current to go through. Third mode is active, where the current flows at a certain rate only, and the fourth mode is reverse active, where the current flows at a certain rate but in reverse direction. These modes, where a circuit is opened closed or closed with resistance, allows for logical calculations and forms the basis of electronics and calculations and computers today. Transistors are used since 1940s, when they replaced vacuum tubes and made integrated circuits possible, which meant a revolution in electronics industry and therefore in our civilization. Also see *Microprocessor, Integrated Circuit, Current*.

UAS:

Acronym for *Unmanned Aerial System*. Please see *Drones and Robotic Vehicles > Unmanned Aerial Vehicles*.

UAV:

Acronym for *Unmanned Aerial Vehicle*. Please see *Drones and Robotic Vehicles > Unmanned Aerial Vehicles*.

Underactuation:

This is the case when the system has less number of motors or actuators than the number of degrees of freedom. The opposite of underactuated is a fully actuated system, where the number of actuation axes by actuators and or motors is equal to the DOF. A fully actuated system does not necessarily mean it is better than underactuated system. In fact, it is the opposite. For example, if we take a few walking steps, but do it very slowly, we have to fully control every movement that our legs make, and this motion does not take the benefit of momentum and inertia of our legs, when they were moving freely in normal walking speed. Here slow walking is an example of full actuation, which is less efficient, and walking at normal speed naturally is an example of underactuation. Study of underactuation in robotics is the key to make robot movements more natural, by taking advantage of natural dynamics.

A walking humanoid robot which does not take advantage of natural dynamics, spends many times more energy to walk the same distance, and at a slower speed, in comparison to a human, due to high joint torque required, to control every move. Robots which use natural dynamics principles are called passive dynamic walkers. These can even walk only by gravity, downward, if designed correctly and in a much more natural way. This is a quite similar principle when flying, for example birds can take advantage of winds, and can fly more efficiently than planes, with a principle called dynamic soaring, as explained in drones chapter of this book. Also see *Passive Dynamics*.

Unmanned:

For *Unmanned Aerial Systems, Unmanned Aerial Vehicles, Unmanned Ground Vehicles, Unmanned Sea Vehicles* please see *Drones and Robotic Vehicles* chapter of this book.

UGV:

Acronym for *Unmanned Ground Vehicle*. Please see *Drones and Robotic Vehicles > Unmanned Ground Vehicles* section.

USV:

Acronym for *Unmanned Sea Vehicle*. Please see *Drones and Robotic Vehicles > Unmanned Sea Vehicles* section.

Vacuum Cleaning Robot:

Please see *Domestic / Service Robots > Vacuum Cleaning Robots* section.

Varistor:

It means variable resistor or also known as voltage dependent resistor, that is used to dampen the voltage variations in a circuit by lowering the resistance when voltage rises and increasing the resistance when voltage drops, so they provide optimal operating conditions and also protect the circuit from the damage that might be caused by voltage fluctuations. When surge happens, they can absorb a high amount of energy, but when inactive, they consume virtually no current. They have non-ohmic characteristics, which means, the voltage-current ($R=V/I$) relationship through it, is not linear. Also see *Current, Circuit*.

Virtual Reality:

It is a computer generated virtual environment in three dimensions, that can allow people to interact with it, using special aids like helmets or 3d glasses. It is also used for robot simulation, and can save significant amount of time during robot development. A good physics engine is required in order to simulate real life conditions as realistically as possible. Also see *Physics Engine, Robotics Simulator*.

Vision Guided Robots:

It is a term used for industrial robots, where the robot operations are guided not only with the programmed motions but also with the help of installed cameras, in order to improve performance and enable them to perform a wider variety of tasks, that would not be possible with programming alone. With the help of cameras guiding the robot, it can identify and track objects, grasp and manipulate them, even if the objects to be manipulated may be in arbitrary configuration. Working in undefined environments is a big challenge for robots and vision guided robotics is an important step to overcome this hurdle. Today there are highly developed vision technologies in the market, that are used by industrial robot manufacturers. For example, spatial vision programs provide real time information about objects in 3 dimensional space, for real time guidance of robots. Other programs are able to identify and inspect parts accurately. These systems must be able to identify and recognize objects in different lighting conditions, random configurations, partially visible situations, at the same time process fast enough to allow the robot work efficiently. Vision guided robotics allow more flexibility, greater inspection and control ability, tracking products more accurately, improved measurement and alignment, improved and wider range of manipulation ability, and therefore it can considerably improve a company's bottom line, because the number of tasks that require manipulating randomly placed objects are far greater than precisely organized objects, which was the only case that robots were able to work in the past, before these vision technologies were available.

Visual Recognition:

It aims to produce meanings out of visual data for image recognition. Based on an image, AI can understand things within a concept such as what is in relation to what and how, by comparing the image with the images it has seen before. By visual recognition, lots of images can be organized and categorized quickly and accurately. It is a sub-field of *Computer Vision*.

Voice Recognition:

This is often confused by speech recognition, however it is different. In speech recognition, the computer understands what is spoken, whereas in voice recognition the question is who is speaking. Here the system requires training, in order to recognize a particular person's voice. Voice recognition systems are much easier to develop than speech recognition systems.

Voltage:

It is defined as the electric potential difference between two points per unit charge. By Ohm's law, it is directly proportional to current and resistance, with the relation $V = I.R$, where V is voltage in volts, I is current in ampere, and R is the resistance in ohms. See *Current, Ohm's Law*.

Voltage Converter:

A voltage converter converts AC and DC voltage into each other. Converters are more efficient than regulators because they bring down the voltage to a higher frequency without wasting any power.

Voltage Regulator:

Voltage regulators generate a regulated voltage in the desired constant amount, higher or lower than the initial voltage. It also takes all the current and creates the wasted heat. There can be step up/down or only step up or step down regulators.

VR:

Acronym for *Virtual Reality*.

Work Envelope:

It is the 3D boundary in space, where a robot arm can reach, when it is stretched at its maximum.

Workcell:

See *Robot Cell*.

Zero Moment Point:

Zero moment point analysis is used, when studying the walking motion of a humanoid robot. For a stable movement of a body, the stabilizing reaction that acts on the body must be adequate both in magnitude and direction, plus it must be concurrent with the moving force, in order not to produce any moment. If not, the moving forces that act on the body will cause it to lose its stability or balance.

For example, if a block of stone is just left in the air, the gravity that acts on it will cause the stone to fall down, because the gravity force is not countered with any reaction force. The same stone, when on the floor, is stable, because the gravity force that acts on the stone is countered by the reaction force that the floor acts on the stone. Similarly, for a moving body, the instantaneous reaction forces that counter the effects of motion must be able to cancel out those effects. If not, the body will accelerate as in the case of falling stone and either fall or lose it balance, until the moving effects are countered again. For example, when you lose your balance for some reason, your body immediately acts automatically, to align your body in the most effective way to counter the moving effects, and that is why you must move your arms and legs automatically, to balance this effect with the help of reaction force from the ground.

Zero moment point, is an instantaneous and constantly changing point while the body moves, where the reaction from the ground, is able to cancel out the moving effects that acts on the body, such as the gravity and inertial horizontal force, by acting in the same magnitude but in exactly opposite direction, as a concurrent force, to the resultant of inertial and gravity force, and therefore it not only cancels out the force, but it also does not produce any moment for the body, since it acts concurrently and in opposite direction to the resultant force, and hence the name zero moment.

If the zero moment point is able to fall within the reach of the legs, the robot (or you) will not fall, by simply moving the leg and foot to the required point (assuming you can move fast enough). If the required zero moment point location is further away, beyond the reach of legs (such as you are pushed too fast from behind), the robot (or you) will fall, such as when you bend your body too much, and your body was excessively horizontal, or you move too fast, and your legs can not stretch far enough, to counter the resultant force of gravity and inertial forces. Therefore, the walking motion of a humanoid robot can be achieved by constantly enabling the zero moment point to fall to required points as required. Also see *Force, Coupled Force, Free Body Diagram, Concurrent Force, Moment.*

ZMP:

Acronym for *Zero Moment Point.*

CHAPTER 2

ANDROIDS / HUMANOIDS

By definition, android and humanoid robots are robots that look like human. Android robots are much more humanlike in appearance and aesthetics, and they may appear almost or completely as human, such as the ones in movies. Humanoid robots however, appear clearly as robots, but they still resemble human shape overall. Therefore the robots listed below are actually humanoids.

To make an android / humanoid robot is as hard as any robotics project can get. In order to give a robot the bipedal movement of two legs, body posture and movement, arm control, hand control, object manipulation skills, navigation, visual, voice, sound and logic processing abilities, all at the same time makes a humanoid robot project really difficult. Almost all the terms and concepts that were explained under Robotics Terms and Concepts chapter for example, are applicable to making a humanoid robot. Even the simplest tasks we do in our daily lives, can prove to be big challenges for a robot, let alone a combination of them, let alone situations that involve decision making <u>and</u> performing those tasks at the same time.

We will eventually see domestic helper humanoids / androids in our homes, but this is not likely to happen before at least a decade. When it happens, it will be one of the things that will change our lives. The impact will be as big as the smartphones had in our lives. We will interact with them, they will help us, indoors or even outdoors, and we will download apps for them, for new abilities, or it will automatically happen as patches, online. They will have a few standard operating platforms, and operating systems, just like different computers by different manufacturers all use one of the few major operating systems. For now, a good candidate for operating system appears to be ROS, Robot Operating System, as discussed under software chapter of this book. And of course, android / humanoid robots will be physically moving entities, and they must and will have safety measures installed and hardwired in them, in order to prevent any harm to a human being or surroundings.

Most humanoids today use motors to move arms and legs and body, but the motors, by their shafts providing torque, do not represent how human body works. The actuation must be done with hydraulic or pneumatic actuators, to represent the linear actuation that our muscles do, and in the long run, with artificial muscles, which have the potential to be hundreds of times stronger than human muscles (see *Artificial Muscles* term, under Robotic Terms and Concepts chapter). Many of the most advanced humanoids robot today use motors. It is true that they are robotics marvels, but with a motorized system, they can only move and carry and manipulate up to a certain limit and cannot benefit from the advantages of linear actuation. A robot that uses hydraulic actuators or artificial muscles, look more natural in many ways, plus they can also behave as underactuated systems, as explained under Robotics Terms and Concepts chapter. Having said that, today the many motorized robots can still perform better than linearly actuated robots.

To come up with the list below, we had to eliminate a lot of old projects. Otherwise, this list could have contained more than 100 robots. And from the remaining, we selected the ones that were either appearing in the news the most and/or the ones that we were aware of and considered the most impressive, high quality and relevant for our book, so here they are... Note that there are also humanoids which are robot kits or toys, and they are presented separately in the next section. For all of the projects or products mentioned in the sections below or in any chapter of this book, it must be remembered that we can only give brief summaries here, our descriptions may not cover all important points, a longer or shorter description here is arbitrary and things we wrote for one robot may also be present in or valid for another, and readers are advised to view a project or product directly from the respective product websites. The robots below are presented in alphabetical order.

Advanced projects

Alter

This robot was developed in Japan in 2016, in Universities of Tokyo and Osaka, and has 42 pneumatic actuators for movement. It learns and acts with its environment through its neural network but the learning and AI seems to be limited, although it can move its arms smoothly. It can also make some facial expressions and it is at research project state. This is not a full humanoid, it is only the body and the arms, but without legs attached.

ASIMO

Created and constantly improved by Honda Motor Company, since 1980s, ASIMO is one of the first robots that comes to anybody's mind in the world, for a humanoid robot. Its frame is made of magnesium alloy, which is covered with plastic, and therefore it is quite lightweight material. Its height of 51 inches (130 cm) allows it to help around the house or communicate directly with a human who is sitting. Its hands are able to sense the force applied and can open its arms by 105 degrees, and with its 14 DOF arms, it can do a wide variety of movements to perform domestic tasks such as pushing carts, carrying handing and receiving trays without spilling the contents, switching lights, performing dance movements, and while doing these, looks very smooth. It can also skillfully walk, turn, run, climb stairs, kick a ball, jump, stand on one foot, with its 12 DOF legs, which brings its total DOF to 57. The 6 axis sensor at its feet receives force input for the robot to respond accordingly. It can also avoid obstacles while moving, with the help of its ground (infrared and laser) sensors at mid height, ultrasonic sensor also at mid height, and visual sensors at the head. The visual sensors are made of two stereoscopic cameras, to judge 3 dimensional distances and objects. It can recognize gestures, faces, voice commands. ASIMO takes part in many worldwide demonstrations. Source: http://asimo.honda.com/

Atlas

Made by Boston Dynamics of USA, it is a humanoid robot with advanced moving skills. It is very strong and fast, with its high strength to weight ratio, as it is actuated hydraulically, instead of motors. It is suited for industrial or even military use, as it can lift loads up to 10 pound with ease. It can walk fast, even on rough or sliding natural ground such as a forest soil covered with ice and snow, rocks, mud. It can also immediately correct itself, even if it starts falling by an external force, such as when someone pushes it hard. And in case it falls, it can stand back up again all by itself. It can open doors, and carry and manipulate objects, lift weights and place them on shelves. It balances itself by using sensors in its body and legs. In addition to its excellent walking abilities, Atlas can also use its arms and legs together for climbing, if the terrain changes from rough, to a climbing surface. Its sensor head includes stereo cameras and a laser range finder. It has a total of 28 degrees of freedom, 1.5 meters tall and weighs 75 kg. Source: https://www.bostondynamics.com/atlas

Durus

Durus is a walking robot, which has remarkable humanlike walking ability, designed by Georgia Tech. University. The robot at this point is developed with the focus of replicating human walking as close as possible and does not include upper body. The robot's feet is shaped like human feet, replicating the curved shape of the feet with arch and heel, which gives it a huge advantage over flat footed humanoid robot legs as far as walking efficiency.

FEDOR

FEDOR, which is the acronym for Final Experimental Demonstration Object Research, is a humanoid robot in development in Russia, by Android Technics company and the Advanced Research Fund, which aims to help astronauts in space or work in extreme conditions. It can turn valves, drive car, use drill, do pushups and a lot more useful tasks such as lifting up to 20 kg, walk, crawl, get into the car all by itself, turn door keys, and use various tools, and seems to be a very advanced humanoid robot overall, as of today. In a demonstration in 2017, the robot was also able to shoot targets by holding a pistol in his hands, although the developers say that it is not designed for any combat operations. The robot is able to act on its own decision in some limited cases. It is expected that by 2021, the robot will be sent to space.

Handle

Created by Boston Dynamics of USA, Handle robot is a very fast and agile robot, which can jump, balance perfectly on two legs, while doing fast turns, can walk on irregular surfaces, climb ramps or even jump from them and not only those but it can carry an impressive payload of 45 kg. It has wheels on its feet, and demonstrates an excellent combination of legs and wheels, benefitting from the advantages of both. The robot is 2 meter tall and weighs 105 kg, and does all these remarkable moves with only 10 joints. Source: https://www.bostondynamics.com/handle

HOVIS Genie

Developed by Dongbu Robot Company of South Korea, HOVIS has a few different models. The HOVIS Eco Plus model is a kit and discussed under Toy Robots / Robot Kits section. HOVIS Genie model is a more advanced model for helping with certain domestic tasks. It also has an on board screen display for interaction. The robot achieves its mobility with a wheel base, can be remote controlled, has speech recognition ability, recorder, speakers and camera and related features to serve as an interactive domestic robot. Source: http://www.dongburobot.com

HRP-4

It is a full size 151 cm high humanoid robot weighing 39 kg, with 34 DOF which allows it to do a wide range of moves. It was made by Kawada Industries of Japan. HRP stands for "Humanoid Robotics Project". The robot also has 5 finger hands for manipulating objects. The first version was HRP 1, which started in 1997, which had used Honda's ASIMO as a starting point, and then the later updates followed as different design. latest update to the robot was HRP-4, which was introduced in 2010. Source: http://global.kawada.jp/mechatronics/hrp4.html

Hubo

Manufactured first by US and South Korean researchers in 2009, under the name Jaemi Hubo, it has gone through several updates since then. The latest version, DRC-Hubo, developed by Korean Institute for Science and Technology (KAIST), can be considered as one of the most advanced humanoid robots in the world today, which also participated in the DARPA grand challenge for humanoid robots in June 2015. The challenge involved tasks such as driving a cart, getting out of it, using drills, turning valves, opening doors, climbing stairs. The robot has wheels on its knees and feet for increased mobility, for saving battery and moving faster on smooth surfaces, while it still can walk as a bipedal robot on rough surfaces or for climbing stairs. On its head it carries lidar and camera. The robot is approximately 160 cm tall and can lift up to 20 kilograms. Its arms have 7 motors each. It has three cameras to obtain a different set of views such as narrow view, wide view or perceiving 3D objects. The robot's hands can exert 200 newtons of force.

Sources:

http://www.drc-hubo.com/

http://hubolab.kaist.ac.kr/p_drchubo

iCub

iCub is an open source and cognitive humanoid robot platform following the GPL license. The robot is an ongoing international research project which is supported by EU Commission under the Cognitive Systems and Robotics program, and Istituto Italiano di Tecnologia (IIT), which is headquartered in Italy and participated by universities from several countries in Europe, such as Germany, France, Spain, Turkey, England, Portugal. The robot was built in the form of a 4 year old human and focuses on learning through interaction with its environment. The project, which started

in 2004, has evolved greatly since then. The robot is 104 cm high and weighs 23 - 25 kg. It has sensors such as 6 Force / Torque sensors in arms and legs, cameras, microphones, gyro and accelerometer in the head, up to 108 tactile sensors in the fingertips and palm, encoders and uses YARP software. It has a tethered power supply which goes through a 48-12 V battery. It has 53 total DOF, 9 in each hand (of which 8 is for thumb, index & middle fingers, which gives the hand dexterity), 6 at head, 7 each arm, 15 at legs and torso together. There are 54 motors controlling 76 joints.

Source : http://www.icub.org/

Photo Credit: Istituto Italiano di Tecnologia (IIT) - www.iit.it

iCub learns through dealing with objects. Photo Credits: Istituto Italiano di Tecnologia (IIT) - www.iit.it

NAO

Developed by Aldebaran Robotics of France, which was later acquired by Softbank Robotics of Japan, NAO was introduced in 2009, and currently has the fifth version as the latest upgrade. It is 58 cm tall and has 25 DOF total. It is an open source robot and used for a lot of research or educational projects too, serve as a companion robot, in addition to being a kit for serious hobbyists. NAO can recognize people, objects with its two high resolution cameras and also has voice recognition. Source: https://www.ald.softbankrobotics.com/en/robots/nao

NAO robot.
Photo by: A. Tuter

Ocean One

It is an underwater humanoid robot developed in Stanford University, Robotics Laboratory. It is a remote controlled robot with force sensors in hands that provide haptic feedback, to give underwater researchers and archeologists the ability to feel where robot touches. The robot successfully completed recovering an artifact from an old shipwreck in Mediterranean. It has two articulated arms, and two cameras for stereoscopic vision. The robot's on board computer ensures that the grabbed objects does not break, and also helps to keep the robot balanced in turbulent waters. Source: http://cs.stanford.edu/group/manips/ocean-one.html

Pepper

It is a humanoid robot developed by Softbank Robotics. The robot does not have legs, it moves around on wheels, so it is only suitable for use on smooth surfaces. It was mainly designed for high level human interaction and it is one of the most advanced in this sense. It can not only recognize speech, but with its advanced cloud backed voice recognition engine, it can also recognize even different tones or nuances in human voice or nonverbal language. With its advanced AI, it can learn from human interactions. Pepper can also use body language effectively, with its flexible very natural looking movements from the waist up, which is made possible by 20 actuators. Source: https://www.ald.softbankrobotics.com/en/robots/pepper

REEM

REEM is a humanoid robot, from waist up and instead of legs it has a mobile base, multimedia touch screen and can interact with (speak to, shake hands, etc) with humans, so it is used mainly for events as a dynamic information point or autonomous guide. Its arms can carry a payload of 1 kg, while its base can carry up to 30 kg. Its two 48 V lithium ion batteries provide power up to 8 hours. Its laser sensor can scan up to 20 meters. For vision, it has stereo and back cameras. The robot's height is 170 cm, and it weighs 100 kg. REEM was developed by PAL Robotics of Spain.
Source: http://pal-robotics.com/en/products/reem/

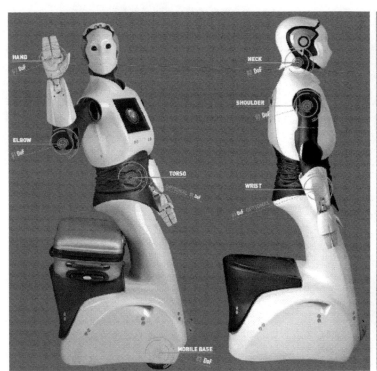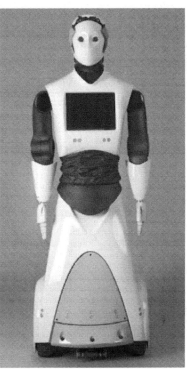

Photo Credits: PAL Robotics - www.pal-robotics.com

REEM - C

Developed by PAL Robotics of Spain, REEM-C is a full size two legged humanoid robot, developed for robotics research. It is open source and upgradable, running fully on ROS. It can lift 1 kg with fully open one arm and 10 kg using both arms. The robot which is 1.65 cm high, and weighs 80 kg, can walk at a speed of 1.5 km / hour, balance itself against pushing, grasp objects, climb stairs and can perform various other tasks. It uses Ubuntu LTS, real time OS, ROS, OROCOS, ROS_control, MoveIt!, walking, grasping, face recognition, speech recognition software. It has a total of 68 DOF, such as 2 head, 7 each arm, 19 each hand, 6 each leg, 2 waist, 3 hip. It includes sensors of stereovision camera in the head, 4 microphones, pressure sensors at hands, 6 force and torque sensors at arms and legs, lasers for navigation, back camera, 5G acceleration IMU with 450 deg/s. The robot has a modular design which is upgradable. It has 3 hours operational time when moving, 6 hours when still. It is also able to speak 9 languages out of the box. REEM-C is suitable for research areas such as navigation, vision, AI, grasping, walking or speech recognition, human robot interaction. It has a simulation model available at: wiki.ros.org/Robots/REEM-C.

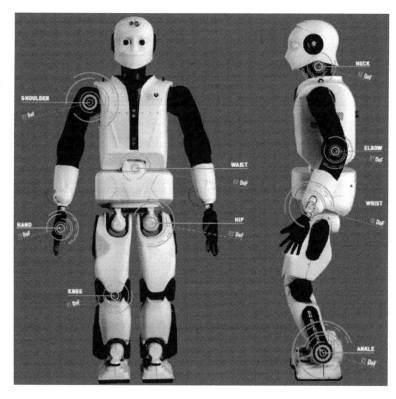

Source: http://pal-robotics.com/en/products/reem-c/

Photo Credit: PAL Robotics
www.pal-robotics.com

Robonaut

Robonaut is a highly dexterous anthropomorphic robot developed at NASA in collaboration with General Motors and assitance from Oceaneering International, to assist humans during space missions, to take on repetitive and dangerous tasks, which enhances the crew's capabilities. It has R1 and the upgrade R2 models, where the upgraded model had significant improvements over R1. Robonaut's hands were designed to provide dextrous and versatile manipulation, in order to use necessary tools, it has advanced vision and image recognition capabilities, and able to respond to unexpected obstacles, which makes it one of the most advanced humanoid robots today.

The robot has a torso with a head, two arms and legs attached but the legs are not designed to walk but rather have end effectors, which includes sensors and cameras and function to enable climbing and movement inside the weightless space station. It is teleoperated remotely to relieve the astronauts from tedious and high risk tasks. It currently operates inside the space station, but in the future it is planned for outside space station and even vehicular use. With its humanlike torso, it can operate like human astronauts. The operator on the ground wears 3D glasses, vest and gloves, and his movements directly transfer to the robot. Source: https://robonaut.jsc.nasa.gov/R2/

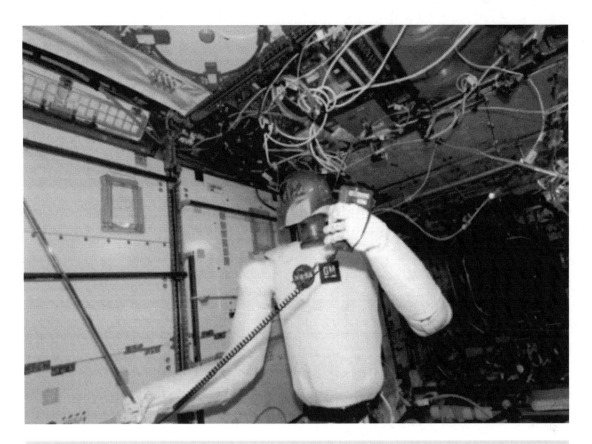

Controlled by teams on Earth, Robonaut 2 measuring air velocity during system check out in the International Space Station. Photo Credit: NASA - www.nasa.gov

Romeo

Manufactured by Aldebaran Robotics of France, which was later acquired by Softbank Robotics of Japan, with collaboration from various French and European robotics institutions, Romeo is a 140 cm tall humanoid robot research project, designed to help people with disabilities and do many different domestic tasks. Source: https://www.ald.softbankrobotics.com/en/robots/romeo

Valkyrie

Built for space exploration by NASA, the project is still under development. It is being designed for future space exploration missions of MARS. The robot, which is also called as R5, has also been given to universities such as MIT and Northeastern University for further development. The robot is one of the most advanced humanoids in the world that is currently under development and is already able to make some smooth natural moves with arms and legs, as well as dealing with objects at the basic level, move around, make autonomous decisions. The arms, legs and hands of the robot has mechanical and electrical disconnects from other parts for easy shipping and service. Some specs about the robot: Height: 6' - 2" (188 cm), DOF: 44 total, weight: 300 lbs (136 kg), battery: 1.8 kWh or tethered also possible. If battery used, it provides 1 hour operation time. Sensors: A perceptual

Image Credit : NASA - www.nasa.gov

Valkyrie Humanoid Robot of NASA

sensor in the head, hazard cameras in the torso. Actuation: 4 series elastic actuators in the forearm, single rotary actuator at the wrist, linear actuators to control wrist pitch and yaw, 3 series elastic rotary actuator for controlling waist rotation and hip rotation joints. Hands: 3 fingers and thumb.

Source:

https://www.nasa.gov/feature/r5

Hobby Humanoid Kits

Other than the humanoid projects presented above, here is a list of other humanoids, which would fall more into hobby category. Needless to say, they all vary in specifications, purpose, skill set, hardware, software and price. The descriptions are very short to fit into pages here. As always, something written about one robot may be valid for others too, so it is advised that readers visit respective product websites. The kits below are all popular ones, with qualities on the high side.

BuzzBot
BuzzBot is one of STEM friendly Jimu robot kits by UBTECH Robotics, which includes pieces to build different types of robots, mostly directed to kids or teens level. https://jimurobots.com/ http://www.ubtrobot.com

HOVIS
Developed by Dongbu Robot Company of South Korea, HOVIS Eco Plus is a 41 cm tall robot with 20 DOF, has non assembled or assembled options. http://www.dongburobot.com

HR-OS
Developed by InterbotiX Labs of USA, this is an open source humanoid platform, which users can build. It has a Linux operated computer and the robot can be heavily customized and modified with 3D components. Users can build programs and apps using different languages to share on tablets, smartphones etc... https://interbotix.com/m/interbotix and http://www.trossenrobotics.com

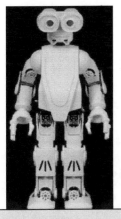

HR - OS5 Robot. Photo Credit: InterbotiX Labs
www.interbotix.com
www.trossenrobotics.com

HR - OS1 Robot. Photo Credit: InterbotiX Labs
www.interbotix.com
www.trossenrobotics.com

Lynx

Manufactured by UBTECH robotics, Lynx is integrated with Amazon's Alexa voice service. It is 17" long and weighs 5 lbs. It can play music, and manage tasks such as scheduling appointments, calendar reminders and emails, it has interaction mode, can serve as video access through camera. http://www.ubtrobot.com

Lynxmotion

Lynxmotion Pete Humanoid Development Platform is a 22 DOF robot, which is also a research platform, that users need to create their own walking algorithm for autonomous or remote controlled operation. It was developed by Lynxmotion, which was later acquired by RobotShop Inc. The robot's parts can also be used to make other robots such as quadrupeds, robot arms, hexapods and more. http://www.robotshop.com/en/

Lynxmotion Pete Humanoid Development Platform
Photo Credit: Robotshop Inc. www.robotshop.com

Meccanoid

Meccanoid 2.0 and Meccanoid XL 2.0 are robotics building platforms that can be built using Meccano parts. Meccano 2.0 is 2 feet tall, with 497 parts, 4 smart servos and 2 motors included, while Meccano XL 2.0 is 4 feet tall, comes with 1014 parts, 8 smart servos and 2 motors. They have both voice recognition abilities and preprogrammed interactive features. http://www.meccano.com

Meccano 2.0 on the left and XL 2.0 on the right. Photo Credit: Meccano Engineering and Robotics - www.meccano.com

ROBOTIS

Below are the humanoid robot kits from ROBOTIS. http://www.robotis.com

ROBOTIS BIOLOID PREMIUM is a modular educational robotic kit with an assembly manual and programming software for 29 different robot examples, which range from a 1 DOF robot to an 18 DOF humanoid robot which is also suitable for advanced robot builders. Photo Credit: ROBOTIS - www.robotis.com

ROBOTIS PREMIUM

ROBOTIS GP (Grand-Prix) is an advanced humanoid robot optimized for various robot competitions. It has high performance actuators and ultra light, high strength aluminum frames. Photo Credit: ROBOTIS - www.robotis.com

ROBOTIS GP

ROBOTIS MINI is a mini humanoid robot comes with an open source embedded board, 3D printable parts, and free Android and Apple Apps. There are various sensors available to expand its capabilities. Photo Credit: ROBOTIS - www.robotis.com

ROBOTIS MINI

ROBOTIS OP2 is a miniature humanoid robot platform with advanced computational power, sophisticated sensors, high payload capacity, and dynamic motion ability to enable research, education and outreach activities. Photo Credit: ROBOTIS - www.robotis.com

ROBOTIS OP
R&D Collaboration with DARWIN Project by NSF

THORMANG 3 is a full-size, open platform humanoid robot with modular design and powerful dual PC-level computing. It has various sensing capabilities and has experience with rescue missions as a participant in the 2015 DRC Finals. Photo Credit: ROBOTIS - www.robotis.com

THORMANG3

Poppy

Poppy is an open source, modular humanoid robot project that was started in 2012. A user community contributes to its development by sharing their work and therefore the project is evolving. The parts are 3D printed by users at their own location which allows of a wider range of possibilities and local creation of customized parts. https://www.poppy-project.org/en/

CHAPTER 3

ARTIFICIAL INTELLIGENCE

Every year, we see increasing number of tasks that can be performed with artificial intelligence. We are approaching to the point where machines will be able to perform the same level of thinking and problem solving skills as human, so that it can practically be considered a conscious entity, such as passing the Turing test, named after the famous mathematician Alan Turing.

As each new task enter into AI teritory, we perceive it as just another automation happening in our daily lives. For example, AI has been checking bank transactions for years now and we are used to it. But virtual assistants in our smartphones or computers today, were considered as science fiction as recently as 10 years ago. Now we are used to that. Same is true for chess playing computers, as we perceived a chess playing AI as science fiction 40 years ago, but by the end of 1990s, AI had beaten the world chess champion Garry Kasparov. Last year, AI system called AlphaGo which was created by the company DeepMind Technologies Inc., beat the world Go champion, which is a game far more sophisticated than chess and we are again perceiving this as just another automation happening. In the games, AlphaGo also made moves that stunned the experts, which were very unusual but far more superior than hundreds of years of human game wisdom would suggest.

There is a point in the future, where AI will match and then exceed human intelligence, which can be a real threat to humanity. Considering exponential growth of technology, we must remember that this exponentiality also applies to the improvement of the AI, and therefore, the developments we saw in the past will be faster always. Based on this, it seems safe to say that this point will be achieved long before the end of the century. For example, due to exponential growth, the knowledge produced, and the technological advancements achieved in just 17 years of this century, is already getting close to the knowledge produced and advances achieved in the entire 20th century.

Different studies come into play when developing AI. Computer science, neuroscience, robotics, even sociology and philosophy. For instance a lot of research is currently underway, in order to study and reverse engineer the brain, for medical purposes as well as developing artificial intelligence. Human brain learns, develops and operates not as an isolated organ but in conjunction with the whole human body, which constantly gets inputs from sensors, not only from the 5 senses, but also senses that processes time, location, motion, emotions and others that we are learning or yet to learn, and because of this, it seems reasonable to say that the AI which will be able to operate and reason like human brain will have to be developed in conjunction with robotics which must involve physical world and physical inputs, and not only by writing code. As an example, a project such as the iCub, which is developed with input from many universities in Europe and located at Istituto Italiano di Tecnologia (IIT) in Italy, follows this route, where a self learning computer works within a human child shaped robot body, constantly getting input from sensors and processing these to learn.

The AI systems can be divided into two categories, in terms of what they are made for, such as the AI developed for making intelligent machines that can imitate human reasoning, and the AI developed for making expert systems that assists humans with intelligent decisions or advice. There are a huge number of applications of AI today, such as in gaming, expert systems, natural language processing, visual processing, intelligent robots that can carry out certain tasks such as assisting humans, automation of manufacturing, building systems, driving and probably hundreds more.

Due to increasing complexity of AI, no job that we humans do today, is safe from AI or automation. Some have already happened, some are happening, and others will happen later. For example, the jobs of factory workers who assembled cars in production lines, are already almost fully automated, while the jobs of construction workers are still not, due to them working in undefined environments with lots of variables from real life, but as AI gets more advanced, even construction is not safe. The more real life variables in undefined structures come into play, the later the automation will be for that job, as in this example. Now let's consider non physical work examples. For instance, the legal search help that lawyers need can be performed to a considerable degree by AI, but the job of an architect or civil engineer is still more intact (except the usage of some programs for complex structural analysis or building design which by the way, get increasingly more complex and capable), although even those are not safe from automation later. For teaching jobs, there are indications that in the near future, learning from AI instead of humans can be mainstream. The examples are in every field to varying degrees. When AI takes over jobs it is not a bad thing. For example, when delivery trucks or taxis start to run autonomously, drivers will lose jobs, but now the costs of delivery and commuting will be lower and overall we will have savings which will have positive impact on our economy so overall more people can be employed, including those drivers who lose jobs, who will be able to find another and higher paying job easier because of bigger economy, when we consider not individual cases but the average. Plus, autonomous cars and trucks will need more people to work on information and computer based jobs, so even the overall number of employed people will probably not decrease, just from this example.

One of the ultimate targets of AI is to give the computer the ability to learn what needs to be learned, in other words, giving it higher goals and let it perform the learning and acting in order to achieve that goal. For example, instead of programming the robot code by code, to tell it how to return to its charging location, telling it to prefer actions so that it can return to charging station. Or, for instance, in a war simulation game, telling it to get rid of the things which are harmful to friendly targets, instead of telling specifically which targets are the enemy.

We can classify the goals of AI into two, as more specific and general goals. Specific or short term goals of AI include, robotics, search and reach solutions, develop strategy against opponents or problem situations, understanding language, understanding images, storing data in a way that they can be processed, learning by examples, generalization or awards, making logical deductions or proofs, dealing with uncertainties, determining specific goals for an individual entity or team. Long term or more general goals include achieving social and general intelligence and even creativity. Probability calculations are used in all fields of Artificial Intelligence. It is used for coping with uncertainties, and chances of occurrences.

There are subcategories in AI, which can be listed as below. Note that these fields are very intertwined, and related, some are subfields of the other, some compliment the others. In other words, all of the terms listed below are subfields of AI, but they are very related to each other, they work together and they are NOT isolated fields at all. These include:

- Deep Learning,
- Machine Learning,
- Reinforcement Learning,
- Self Learning,
- Data Mining,
- Computer Vision,
- Machine Vision,
- Object Recognition,
- Pattern Recognition,
- Visual Recognition,
- Image Recognition and Processing,
- Cognitive Computing,
- Language Perception,
- Expert Systems,
- Neural Networks,
- Deep Neural Networks,
- Natural Language Processing

All of these AI subcategories you see above are listed and explained under *Robotics Terms and Concepts* chapter, which is the first chapter of this book.

Recent AI Projects and Advancements:

In order to see the big picture, let's talk a little bit about some of the AI projects that made the headlines in the recent years. Due to the exponential increase in development speed of AI, we see new projects more and more frequently. It is starting to seem almost routine, to see a headline every week that says for example "X company started to work on an autonomous navigation system" or "A startup company has found a new way to language processing from videos". So the ones below are just a few of some well known examples in a growing list of successful or promising AI projects, with different goals, uses and approaches.

DeepMind Technologies Inc.
No doubt, many people felt that they witnessed a historical moment, when they watched the world Go champion Lee Sedol, resigned to its AI opponent, AlphaGo last year, in 2016. The program, which beat the world champion 4 out of 5 times, was made by a UK based company DeepMind

Technologies Inc., which was started in 2010 and later was acquired by Alphabet Inc., the parent company of Google, in 2014. In addition to studying millions of positions, which the company said was equivalent to 80 years of playing Go, the program made its success also by the deep neural networking, which enabled it to perform reinforced learning, which learns just from raw data and experience. Go was viewed as a big challenge for AI, because it is a game that requires intuition and only making calculations like in chess would not solve the problem. Readers can view the match and more on AlphaGo from DeepMind website here: https://deepmind.com/research/alphago/

One common problem of neural networks in general is that the knowledge acquired for a task is immediately forgotten, after the system starts working on a new task. DeepMind researchers also worked on this problem and came up with a solution, inspired by neuroscience, which they call Elastic Weight Consolidation EWS. The name comes from the process of modeling the neural connections as elastic springs, where, the spring gets stronger as the importance of the previously learned connection increases, and thus the overwriting of that connection becomes harder, which in turn enables "remembering" of the previously acquired neural connection or in other words protecting it. DeepMind researchers published this work on a paper, which can be viewed here: http://www.pnas.org/content/early/2017/03/13/1611835114.full

DeepMind recently came up with an open source development platform, DeepMind Lab, where there are 3D simulated environments similar to computer games, and it is viewed through the eyes of a moving sphere called "the agent". The agent levitates and moves around and performs various tasks in a rich 3D environment full of simulated objects, obstacles and challenges. All these facilitate the development of AI in terms of navigation, motor control, strategy, 3D planning, memory, where fully autonomous agents develop themselves by self learning. The idea here is that the development of AI cannot be isolated from operating in a physical world, just like we humans develop our intelligence, through the use of our bodies, interacting with the environment and objects, manipulating them, using our senses, also interacting with others, either physically or verbally. Note that each of these concepts represent a huge area to be worked on just by itself, let alone the combination of them, and therefore, open source community contribution seems to be a good idea here. More details can be seen here: https://deepmind.com/blog/open-sourcing-deepmind-lab/

DeepMind also has a lot of publications, which can be found here: https://deepmind.com/research/publications/

Google - Image Captioning AI

In September 2016, Google demonstrated a program that can caption images with high accuracy. The image captioning not only identifies objects in the picture, but also can describe what they are doing. Google made it open source in order to get contributions from community. The paper released by Google can be reached here: https://research.google.com/pubs/archive/43274.pdf

Thinking about the near future, image captioning and describing its contents can have a big positive push to developing AI for robots that will help us around, because with this technology, if and when an AI is able to describe its environment accurately, it can also take actions about it. Or thinking in reverse, when it receives a command from a human in the form of a sentence in natural language, it can make up a picture in its mind based on the described situation and act accordingly. Of course to write "act accordingly" is easy here, and appears natural to us humans, but for robotics, it means years of research and development, which will enable a robot to move its limbs in coordination with the ultimate goal of achieving something "similar" to that target picture or situation, after comparing it with the current picture, which seems not very different than how we would do it anyway, as humans.

IBM - Watson

Several years ago, an AI system called Watson, created by IBM Corporation, made the headlines, when it was able to defeat human champions in the popular TV show called Jeopardy!, made by Jeopardy Productions Inc., where questions are asked in everyday language to competitors. Watson is a system that is able to answer questions in natural language. It is able to access vast amounts of data, when answering questions, and this makes it a very useful tool for many applications. Watson is now being used in healthcare for example, in diagnostics, drug discovery, genomics and several other areas such as in tax preparation, chatbots, weather forecasting. Main website: https://www.ibm.com/watson/

Watson also has a developer cloud, which allows development in many fields such as visual recognition, natural language processing, tradeoff analytics, translation and more. It can be accessed here: https://www.ibm.com/watson/developercloud/

Quantum Computing

Although quantum computing is a subject of computer engineering rather than AI, it is still worth mentioning here, because of the game changing technology it will create for computing. The computers that we use today, make their calculations by using on or off position of individual transistors, in other words, with 1s or 0s and the logic gates produced by them. No matter how complicated software or algorithms you used in your life so far, they all come down to 1s and 0s. A transistor can either be on, such as 1, or it can be off, such as 0. The tiniest piece of information which stores this 1 or 0 is called a bit. Today, we are able to squeeze billions of transistors into a square inch but it seems that we are approaching a limit for this.

Quantum computers do not use bits, but they use quantum bits or in short, qubits. A qubit is a state of an atom, that can have either 1 or 0 or both 1 and 0 at the same time, instead of only 1 or 0 as in the case of ordinary bits. This is based on the quantum effect called superposition. Using this principle, it is possible to make exponentially stronger computers than ordinary computers that we use today. Consider the case of 1 qubit, which can be 1 or 0 or both. To represent this situation that we represented with 1 qubit, we need 2 ordinary bits. Now consider 2 qubits and you have $2^2 = 4$ possibilities. This 4 possibility can be represented with 4 ordinary bits. Now consider 3 qubits and you have $2^3 = 8$ possible calculations, which is possible to show with 8 ordinary bits. Next, 4 qubit will be equal to $2^4 = 16$ bits, 5 qubits will be $2^5 = 32$ ordinary bits. So, for example a computer of 100 qubits, will be equal to a computer of 2^{100} bits as we use today, which is really, really a big number. As you can see, each time you increase qubit by only 1, you exponentially increase the possibilities that you can calculate by 2. This is what gives exponentially greater power to quantum computers, over the computers that we use today. Quantum computers are probably showing sufficient promise today, that in March of 2017, IBM announced that it will start a new division called IBM Q, to start working on quantum computers. Although building actual quantum computers may not be possible for at least several years or more, the research is underway.

The main challenge in front of building these computers is that, the effect that we just described above, in other words, the atom being a 1 or 0 or both, dissolves as soon as the effect is measured and the information is lost. In short, what scientists are working on is to make this measurement in such a way that it will not disturb the quantum state and obtain information as well as making calculations with it.

CHAPTER 4

DRONES & ROBOTIC VEHICLES

First let's define drones clearly. Drones are mobile robotic devices, that are either remote controlled by humans or move autonomously. Almost everyone, immediately thinks that drones represent only flying unmanned vehicles, such as unmanned planes or multirotors, but technically it is not true. Unmanned aerial vehicles, (UAV), unmanned ground vehicles (UGV) or unmanned sea or underwater vehicles (USV) are all drones and they are all separate sections of this chapter. The main thing is that the drones are robots that <u>move</u> and they are <u>unmanned</u>. Although technically this is a very wide definition, the term robot is still used more often, even though many of them will also fit into drone definition. Below we discuss aerial, ground and sea drones.

Unmanned Aerial Vehicles (UAV)

This section is only about flying drones, and for convenience, when we say "drone", from this point forward in this section, it will mean flying drone only. Flying drones are also called UAV, which stands for unmanned aerial vehicles or UA, unmanned aircraft, or UAS, unmanned aircraft system.

Drone use mostly started more than 10 years ago in military at first, but as the battery and automatic pilot technologies improve and decrease in cost, we begin to see drones increasingly in our daily lives, with longer flight times, faster speeds, better control features, higher payload capacities, more capable autonomous features, which means more frequent use for more number of tasks.

Types of UAV

Planes: Unmanned planes follow the same main principle as manned planes, which is based on lifting the aircraft by means of pressure differential on different sides of the wings, when a propeller in the front pushes the air towards the wings. Planes fly forward faster than helicopters or multirotor drones but they cannot hover in the air like them and do not have the ability to takeoff or land vertically.

Helicopters: Just like full size helicopters, unmanned helicopters also have two rotors. First is the main bigger rotor that pushes the air downwards to do the lifting, and move the helicopter forward, and the smaller rotor attached to the tail controls which direction the helicopter goes. The small propeller at the tail also counteracts the rotation effect that the main propeller creates. Similar to multirotors, helicopters also use gyroscopes in order to keep balance against the effects of wind, or the turning effect which is a result of throttling the helicopter up or down.

Multi Rotors: This is the most popular type of drone in use today in general consumer and hobbyist market. The multirotor system offers an important advantage in comparison to single rotor helicopters, which is the greater level of control and also the ability to stay in the air even if one of the rotors fail. More rotors of course allow greater carrying capacities, and mean smaller rotor blades which are easier to use, but stronger and heavier battery is needed to run more blades. The most common type of multirotor drone is a quadcopter, which means, a drone with 4 blades. For heavier loads 6 (hexacopter) or even 8 (octacopter) propellers may be used.

Current Uses of UAV

- Military
  Examples: Reconnaissance or attack drones.

- Visual inspection, photography or video recording
  Examples: Wildlife control, ski, fire, border patrol, stadium recording, surveying and mapping, traffic control, photo taking, film recording, construction or existing structures inspection and many more...

- Delivery
  Examples: Large online retail companies are now seriously considering delivery by drones but trying to overcome obstacles about regulations. The delivery can be achieved not only by UAV, but also with a combination of UGV and UAV, such as the UGV driving on land and the UAV doing the required delivery as needed, when a target delivery location is close, and then returning to UGV for recharge and pickup the next item for next delivery. Drone delivery may be common in the near future.

- Rescue and emergency deliveries
  Examples: Delivery of emergency supplies to mountaineers, rushing life vest to people in the sea, quickly searching an area for survivors after disasters

- Law enforcement
  Examples: Monitoring an area flexibly, safely and remotely, following of escaping suspects, safely delivering or removing something to / from a specific location

- Broadcasting
  Examples: Beaming down internet through the use of drones.

Terms in relation to UAV, Aerial Drones

Accelerometer:
See *Sensors* chapter.

AHRS:
See *Attitude and Heading Reference System*.

Aileron:
See *Roll*.

Airspeed Sensor:
It measures the speed of the drone relative to the air, by measuring the positive and negative pressure differences around the drone. When purchased, they usually come together with pitot tube and connection cables. It is recommended for advanced users or drones only, as it necessitates an extra layer of control and tuning. Through pitot tube, the pressure is measured and then this is converted to air speed. Air speed varies with the square root of air pressure. The pitot tube, which takes in the air, transmits it to the sensor through rubber tubing. The sensor is connected to flight controller through a 4 wire I^2C cable. Air speed of drone is different than its speed relative to ground. When calculating flight time for a certain distance, the ground speed is used. For example, if the aircraft is moving in the air with 200 km/h, into a headwind of 5 km/h, then its ground speed is 195 km/h. This is how fast the shadow of the aircraft moves on the ground. When airspeed is corrected for pressure and temperature, true airspeed is obtained. This is the true speed at which the aircraft moves through the air fluid that surrounds it.

Altimeter:
See *Sensors* chapter.

Attitude and Heading Reference System:
Also known as AHRS in short, it provides heading and attitude information, with a 3 axes sensor system in 3D space in real time, for heading, roll, pitch. They work by combining gyroscope, accelerometer and magnetometer data, all are in 3 axes, with complex algorithms such as Kalman filtering, and can provide better results than systems based on only gyroscopes. Gyroscopes tend to drift, and in an AHRS, drift from gyroscopes are compensated by gravity and earth magnetic field vectors, which provide absolute references for gravity (vertical) and magnetic field (horizontal), which are absolute down and north references respectively. AHRS costs and sizes now dropped significantly and they are now within the reach of everyone and can be as small as a coin. There are many applications of AHRS, not just for unmanned vehicles but also for applications such as analyzing human motion, hand held devices, camera stabilization systems.

Autopilot:
See *Flight Controller*.

Bank Angle:

This is the angle between the wings of the aircraft and the horizontal plane, when the it flies inclined. For example, when the drone flies in perfectly horizontal position, the bank angle is zero, and when it starts to tilt in one direction, bank angle increases.

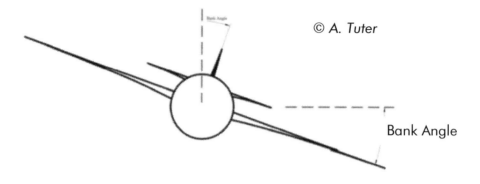

© A. Tuter

Bank Angle

Battery:

Batteries provide essential power to the motors, receivers and controllers of a drone. For multirotors, the most commonly used batteries are lithium polymer (LiPo) types, as their energy efficiency is high. Usually 3-4 cell batteries are used, which provide currents of up to around 5000 mah (miliamperes - hour) capacity. To understand what mah means, consider this example: a 3000 mah battery will last 3 times longer than a 1000 mah battery. As an example, think of the charge (or load) in amperes and time as similar to velocity and time relation, for convenience. velocity x time = distance. Here the distance is mah, so in other words, it is the distance you can go for so many hours at a certain speed. As the speed (ampere or load you use) increases, the time will decrease because you have a certain defined mah limit (distance). The advantage of LiPo batteries are that they can discharge at a much faster rate than a normal battery. It is recommended to buy a few sets of batteries, so that, when the first set is discharged, you do not have to wait for flying your drone again, and while one charges in the recharger, you can use the other battery. Some intelligent batteries on newer models have sensors and they can calculate its distance from the base versus amount of power to return. **Safety Note**: Lithium batteries can catch fire. Check safety requirements of the battery manufacturer. Batteries are also explained in a more general sense under *Robotics Terms and Concepts > Battery*.

LiPo Battery. Photo Credit: Hobbico - www.hobbico.com

Catapult Launching:

This is one of the methods to launch airplane drones, because airplanes need initial speed in order to fly. Catapults are used in order to throw airplanes into the air easily and quickly, where there might not be enough distance to speed up, or the drone might not have the gear to speed up (which saves weight and control systems). Catapult launched airplanes will need additional reinforcement in order to withstand the throwing force from catapult.

Photo Credits: UAV Factory - www.uavfactory.com

Dead Reckoning:

To estimate the position only by using data from internal sensors, rather than GPS, relying on a previously known position. This is useful when there is interference with receiving data from GPS, such as indoors or near high structures or tunnels. Based on the last known position of the aircraft, (this could also be a land or sea vehicle), its position can be calculated only by knowing the distance and direction travelled since that last known point, which is measured by inertial navigation system, which include accelerometer, gyroscope. Because of relying on only one previously known point, dead reckoning calculations are subject to cumulative errors.

DGPS (Differential Global Positioning System):

This is an enhancement to GPS, which provides improved accuracy, down to several centimeters, in comparison to meters for regular GPS. DGPS uses a local base station for reference, to enhance accuracy. The signals from base station is received by a GPS receiver on the drone. DGPS provides differential corrections to a GPS receiver. This improves accuracy and monitors integrity of information sent by GPS satellites.

Drift:

Drone moving in a direction other than intended. Drift may be caused by winds or sensors that are not adjusted well.

Dynamic Soaring:

A term used for taking advantage of winds, while flying, to reduce energy consumption. Currently planes are far behind nature on this, where birds can very effectively take advantage of wind energy and fly for long periods of time, with little flapping of wings. For example an albatros bird can stay over the ocean for days or weeks, with hardly flapping its wings, using this principle. The wind speed changes at different heights above the water and it decreases the closer it gets to the water because of friction with sea surface, and directly above the water it reduces to almost zero. The bird uses the wind to climb, makes a curve at a certain height, descend, and when it reaches almost to the surface, it again orients its direction to take wind and climb and so on, and can travel long distances by drawing curves this way. The potential plus kinetic energy of the bird is constant and "refilled" constantly by wind energy, which is just enough to overcome drag from wind. Reproducing this flight ability for drones means they will fly much longer durations and distances with less fuel. Also see *Thermal Soaring, Ridge Soaring.*

Elevator:

See *Pitch.*

Flight Controller:

Also called the autopilot, or control board, this is the brain of the drone. It is an integrated circuit, which includes sensors, microprocessors, and input output pins. Flight controller not only directs drone where and how to fly, but it also provides stabilizing, hovering ability in the air, by counterbalancing the effects of wind. Levels of stabilization of drone can also be adjusted in some models by limiting the bank angle to high or low levels.

Advanced flight controllers can:

- Start hovering instantly in the air as soon as you release controls
- Autonomously take the drone from one point to another
- Return the drone to the base location in the event of a problem or command from the base
- Follow a moving person or object autonomously at desired altitude, distance, bearing, angle
- Orbit an object
- Target a certain location and altitude
- Avoid obstacles in the air autonomously, indoors or outdoors, by also using cameras and ultrasonic sensors
- Focus camera gimbal at a point of interest, while the operator directs the drone
- Perform waypoint navigation, where the drone can travel along specified waypoints for a desired travel path
- Take panoramic photos of a target by directing the flight controls and camera gimbal accordingly

Although we focus our attention on drones here, it must be noted that autopilots can also guide manned aircraft. These can release pilots of tedious high altitude cruising tasks. Some can even do very precise maneuvers, such as landing an aircraft in zero visibility weather. Autopilots, when operating totally autonomously, are also able to control drones much faster and accurate than human operators.

The chart below summarizes the direction of flow of commands in a drone. It can easily be seen that the flight controller is the center of everything:

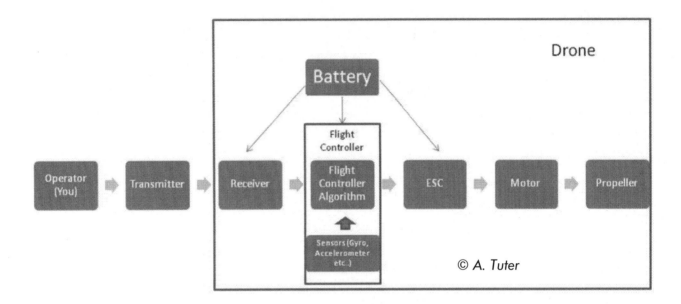

© A. Tuter

Flight controller receives information from two sources: First, it receives commands from the transmitter on the ground that is used by the drone operator, and second, it constantly receives feedback from gyro and accelerometer (which is in INS), plus other sensors if any, that it contains. With the help of INS, the autopilot constantly manages the hovering, tilting and speed of the drone as required. Other sensors may include barometer, which helps autopilot to keep altitude still or be adjusted as required, air speed sensor, ultrasonic sensor, which serves like a radar by reflecting sound waves on surrounding surfaces especially indoors to determine position, magnetometer, which measures magnetic field, GPS sensors, which helps to position the drone whenever satellite data is not obstructed.

After the combination of these two main sources of information are processed, organized into useful signals by using attitude estimation algorithms to communicate to multiple ESCs, the resulting commands are sent to each ESC at the right moment in a very fast ongoing process, in order to adjust the turn of the motors via signals. The signal from the flight controller is usually a PWM (pulse-width-modulated) waveform although other types of signal can also be used with some ESCs.

There are different types of flight controllers, for different styles of flight. The three different styles of flight are sports flying, aerial video recording and autonomous flying. Each of these require different set of functions from an autopilot. For example, if the purpose is to record videos, the flight controller must be capable of providing a very smooth flight. For autonomous flight style, the autopilot must be capable of performing relatively complex autonomous tasks. For the sports or racing mode, he flight controller must be able to handle high speed maneuvers better. Still it must not be forgotten that, flight controllers are configurable, and one flight controller may be set up to different configurations for different flight missions.

As a note for D-I-Y hobbyists, building a system to do all these

Example of a flight controller. Photo Credit: Radiolink - www.radiolink.com.cn

is very difficult even for a person that has years of experience in programming, flight and electronics. If you are interested in developing or participating in building autopilots, we suggest that you participate in one of the open source autopilot projects.

Examples of such projects and code repositories are:

ardupilot.org   autoquad.org   dronecode.org
openpilot.org   pixhawk.org   px4.io
smaccmpilot.org   github.com/multiwii

Flight Logs:

It is the form that is filled by the pilot, to keep a record of the flight information. Some advanced drone models can log and remember the flight details and do this automatically, by recording flight data from internal sensors such as flight distance, time, location, route can be logged in using GPS sensors. These and many other information about flight can be entered manually to keep logs. Some companies offer automated fleet management programs to track data.

Flight Time:

It is the total time that a drone can spend in the air. Most available and popular drones for general consumer market have flight time range anywhere between 10-30 minutes. The drone operator must therefore carefully estimate the furthest distance that the drone can travel before losing contact and having enough time to return to the start point. Some drones can automatically estimate this based on their GPS sensor information. When making flight time calculations, the ascend and descend rate of drones must also be taken into account. Note that for a tethered drone, the flight time could be hours or days, as the power can be transmitted through cable.

FPV Flying:

Acronym for first person view flying, it means to have an on board camera that enables to see things that are captured by the drone camera in real time. It allows better, more entertaining and immersive control of the drone, with respect to hand control and therefore especially preferred by drone racers, who must have the fastest and most accurate control of the drone at all times. Although it provides better control of the drone, it must not be forgotten that the operator is only limited to the camera view here. The rule of operating drone within line of sight is of course still valid when doing FPV flying, as if doing normal control without FPV. It is recommended to learn hand controlling a drone with a cheaper model first and then move on to FPV control, which is more expensive.

FPV Monitor. Photo Credit: GetFPV - www.getfpv.com

Frame:

This is the actual body of a drone, the skeleton that holds all the electronics parts, motors and propellers together. It can be of plastic, PCB, aluminum, carbon fiber, foam or even wood. It must be of strong and light material, rigid enough but also crash resistant. It must allow proper distribution of weights of components of the drone. Carbon fiber is a tough and lightweight material and therefore suitable however, it impedes the radio signals. Wood has the advantage

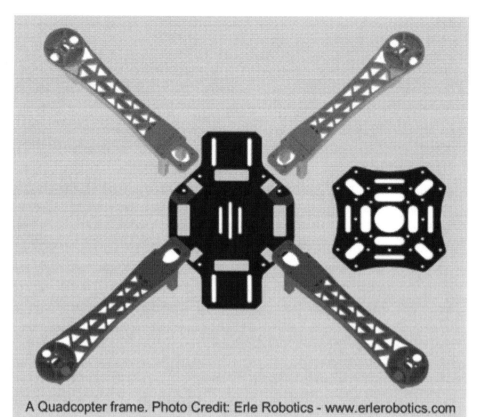

A Quadcopter frame. Photo Credit: Erle Robotics - www.erlerobotics.com

of being cheap and it can easily be replaced. Aluminum is also easily replaced, inexpensive and easily workable. Some frames are foldable, for ease of transporting. Frames with longer arms cause higher moment of inertia when controlling the drone. The most popular frame for multirotors is a quadcopter frame, which is a frame with 4 arms. The arms can be separate than the main body, which allows easier replacement or be manufactured together with the main body.

A foldable frame. Photo Credit: Rosewhite - www.rosewhite.de

Fuselage:

The main body of an aircraft where wings and everything else is attached and placed into.

GCS:

Acronym for ground control station, GCS are used to control advanced grade drones, such as military or mapping drones, by the operator from the ground. In addition to controls in a transmitter, these also have a laptop and special software to perform different operations for control and information processing such as maps, target detection and more functionalities with respect to a transmitter. Unlike transmitters, ground control stations may have unlimited range. GCS can control more than one UAV. Below you can see an example.

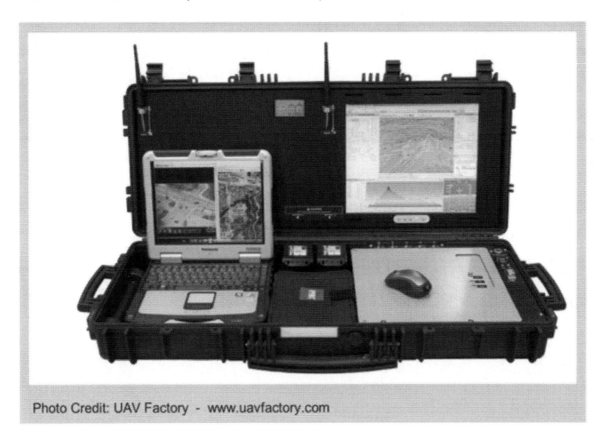

Photo Credit: UAV Factory - www.uavfactory.com

Geofence:

A geofence is a virtual barrier that the drone must observe, for a real geographic area. It is imposed by the program as a radius from a certain start point, or, could be defined manually preflight, such as setting airports or school areas as no fly zones, in order to keep the drone within those boundaries defined by the geofence, and prevent it from entering into no fly zones.

Geospatial Data:

Data that has geographical component. For example, the data that is collected by drones from the surface of earth for mapping operations is geospatial data.

Gimbals:

These hold the cameras in place and attached to frame of the drone and used to keep the camera stable in order to take smooth videos and pictures. They are also used to control the direction of cameras. Gimbals are 3 axes which means it can direct camera or counterbalance the negative movement effects of wind or shaky movements you cause in all axes. Even if the drone shakes considerably, gimbals can ensure that the camera stays at level surface, as the sensors inside can correct the orientation many times per second. Therefore gimbals also have motors to control and counter all these movements. If you only want to take photos and not videos, you may not need a gimbal, and eliminate the extra weight and cost. In different flight modes, the gimbal is controlled by autopilot. See the photo under the term *Landing Gear*.

Global Positioning System:

It is a navigation system, where, location and time information about anywhere on the planet can be seen, as soon as the receiving device sees at least four satellites. On drones, the GPS chip communicates the location of the drone to controller and serves as one of the sensors of the drone. This chip also records the drone's starting position, and can enable flight controller to return the drone to that start position autonomously. This chip is also necessary to automatically hold the drone in stable position horizontally, such as adjusting its position automatically to compensate movements caused by wind, when the operator wants to hold it in stable location.

GPS:

See *Global Positioning System*.

Ground Control Station:

See GCS.

Gyroscope:

See *Sensors* section.

IMU:

Inertial measurement units are measuring devices that act like a combination of accelerometer and a gyroscope, in order to give complete information about drone's position, orientation, speed, acceleration. IMUs can also contain magnetometer, in order to also give information about magnetic field. By using IMUs, the drone can also track its position only depending on sensor input from accelerometer and gyroscope, without the need of GPS. This is an especially useful concept where GPS signals are not available, which is known as dead reckoning. As dead reckoning is subject to cumulative errors, the sensitivity of measurements provided by IMU is important.

INS:

Acronym for inertial navigation system. It is a system that calculates position without the need for external position aid, such as GPS. It contains IMU. Also see *Sensors* chapter.

Initializing:

This is done to initialize the gyroscope, before the flight, by placing the drone on a level surface and letting the gyro "understand" the level surface.

KV Rating:

This is used to rate the motors, where motors are identified by their KV rating, which is a value that shows how fast a motor will rotate under a certain voltage. For a multirotor, a KV rating of up to 1000 can be suitable, whereas for more aggressive flight, such as action recording or acrobatic flights this value can be more. For example if a motor has a KV rating of 1000 rpm/V, then, under 7.4 V, the motor would rotate 1000x7.4 = 7400 rpm.

Landing:

Multirotor or helicopter drones land and takeoff vertically, but planes land while decelerating and moving forward. For this reason, the landing gear of multirotors and helicopters are in the form of skids, while planes land on wheels.

The time for landing must be taken into account when calculating flight times. Landing takes place slower than ascending, due to the precise maneuvers needed.

Also, helicopter or multirotor drones must land slowly, or they will get caught in the downward pushing air that their own propellers generate. If the drone gets caught in this downward pushing air, which sucks the drone downward like vacuum, trying to throttle up to escape this will cause even a stronger vacuum, pulling the drone down even more.

Landing Gear:

For multirotors, these are bent shaped skids that protect drone in case of rough landings. They come in different shapes and sizes depending on the need. The main idea is to absorb the shock of rough landing so as to minimize the damage to the drone. Some landing gears are fixed and others are retractable in order to allow the camera to turn and also not to block the view when recording videos or taking photos. They may be of carbon, aluminum or special plastic material. Retractable gears come with servos for retraction. While fixed landing gears are very cheap, retractable ones may considerably be more expensive.

Landing gear and gimbal example.
Photo Credit : Erle Robotics - www.erlerobotics.com

LED Lights:

LED lights are sometimes used on drones for illumination of the drone or its surroundings, as well as aesthetic reasons during night time flights. It is also required by many drone race organizers.

Example of LED lights and LED lights attached to frame. Also see the PDB in the middle, for the lights. Photo Credit: GetFPV www.getfpv.com

LIDAR:

See *Sensors* chapter.

Lift:

The resultant upwards force created by the propellers in order to lift the aircraft. This is achieved either through creating pressure differential at the faces of the wings in case of a plane, or pushing the air down in case of a helicopter or multirotor. Number of propellers, size of motors, battery power all affect the lift force.

Motors:

Motors are powered by the batteries and make the propellers turn. Motors in a multirotor drone do not draw power directly through the battery. The power is transmitted through ESC, for each motor, and the ESCs get their power from the power distribution board, which is powered by the battery. Larger drones need larger motors and larger batteries. Motors are identified by their KV rating and thrust. Motors were also explained in a more general sense under *Robotics Terms and Concepts* section.

Mounting Plate:

This is the center piece of a multirotor frame, that holds the flight controller, battery, receiver, camera. Around the mounting plate, the arms of the multirotor are attached. The term mounting plate is also used for the pieces that attach motors to the end of the arms of the drone.

Operational Range:

How far the drone can operate safely, measured in horizontal, without losing contact with the transmitter, and allow the drone to return starting point without running out of batteries or fuel.

Parachute:

Using a parachute adds to the payload but it is a good idea in order to prevent crashes in case of a malfunction or other unforeseen circumstance. The parachute slows down the fall and prevents damaging crashes which may damage the drone itself, property or people.

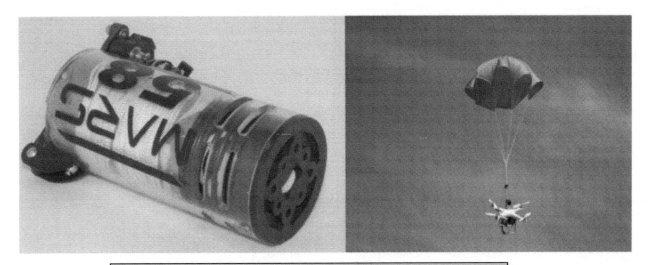

Photo Credit: Mars Parachutes - www.marsparachutes.com

Payload:

It is a term to express how much weight the drone can carry, in addition to its own weight. The larger the propellers of a multirotor drone, the more payload capacity the drone has. But it must not be forgotten that larger propeller can only be turned efficiently with larger and heavier motors and batteries. The total weight of a drone is one of the biggest factors effecting its performance, as far as the flight time or maneuverability. For example a reduction of %20 of total weight of a drone's weight, may mean a %20 increase in flight time, given the same battery capacity, propeller and motor. Therefore, the payload must be chosen carefully. Heavy lifting drone frames must especially be of durable and strong materials due to increased loads taken during maneuvers. Most multirotors in general consumer market today can lift anywhere between 1-10 kg of payload, which also affect their flight durations.

Phone Apps:

Many drone manufacturers have come up with apps that make control of drones via smart phones. Through these apps,

- the drone can be controlled just by tilting the phone or touch screen commands
- setting of many functions for autopilot for flight modes can be made
- the camera and gimbal can be controlled
- real time, FPV viewing of camera o screen and sharing photos or videos instantly is possible
- real time flight parameters can be seen on screen
- detailed records of flight can be kept
- simulated flights can be flown
- many other functions not listed here, depending on the manufacturer

PID Settings:

PID stands for proportional, integral and derivative. It is used in industry to control and correct processes based on the error values obtained, acting like a constant feedback loop (or control loop). For drones, this principal is used for correcting how the drone flies when it tilts by an external force such as wind, or reacts too much to the operator's move commands or the center of gravity is off. PID settings should be tuned before flight. It is done as a trial and error process when you tune them for your drone. For example if in the beginning the P setting is too low, the multirotor will roll, yaw and pitch a lot and will be slow to react. It should be increased until it becomes stable. However turning it up after a critical point will make it oscillate with high frequency, so if you want a smooth flight, refrain from increasing this value too much. P setting is the main value. I setting is to overcome external effects like correcting for wind. If I value is too high, the drone will be hard to control and rotate in all axes. And finally D setting is for controlling the response of the drone when you make aggressive moves such as action recording, so for these kind of flights, D setting should be high. In other words, this will increase how fast the drone reacts to the user. P is for present errors, I is for the total of past errors and D is for the estimated future errors. PID controls work by measuring the actual angle feedback from the gyro sensors and comparing it to the desired angle, for all three axes of turning. So this is the core software algorithm of the flight controller.

Pitch:

Tilting the multirotor forward or backward. It is also called elevator. This is achieved by turning blades faster on one side and slowing down at the other. (For the pitch of propeller, see propeller). The figure below shows the turning for *Pitch, Roll* and *Yaw.*

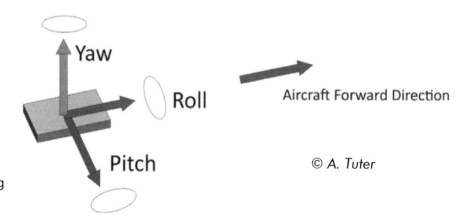

© A. Tuter

Power Distribution Board:

These flat boards contain soldering points for cables in order to distribute power within the drone between different electronics. PDBs have input and output points, or positive and negative terminals which are neatly connected to each other. Some drone frames may already have power distribution board build into them except carbon fiber frames, because carbon fiber conducts electricity and the board must be mounted separately in this case in order to avoid short circuits. Simply by connecting the cables, without a PDB, power can also be distributed, but the board helps keep everything tidy. PDBs come in different sizes and current ratings, and different number of connection points, and some include BECs, (voltage regulators). For very high currents PDBs are not suitable but for a normal hobby grade drone, it is unlikely that there will be currents that PDBs cannot handle. Also see *Circuit Board*, which is a more general term, to include any type of circuit on it, in addition to just power circuit, under *Robotics Terms and Concepts > Circuit Board*

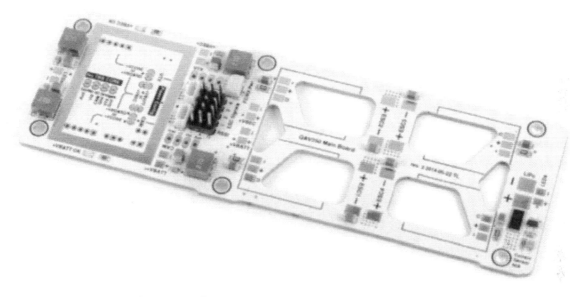

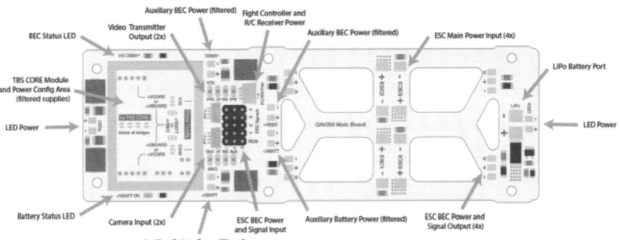

Power Distribution Board Photos By: GetFPV - www.getfpv.com

Propellers:

Propellers carry the drone by turning and pushing air downwards. They are delivered in clockwise and counterclockwise turning pairs. In other words, both pairs push the air downwards, but one pair turns CW and the other turns CCW. For a multirotor, buying CW and CCW propellers in different colors can help identifying them quickly on the drone in order to determine the orientation of the drone, because it is important to know the front side of the multirotor during flight and from ground. The frame colors will also help. Turning of propellers in CW and CCW direction enables rotating or tilting the drone, as explained by the graphics under the term pitch. In case of a plane, the propellers push the air backward to underside and upside faces of the wings so that the upper face has small pressure and lower face has high pressure which lifts the aircraft. A certain size propeller has a certain lifting capacity and beyond that, it cannot push the air efficiently. Propellers must be chosen for their size and pitch. The pitch of propeller is the angle which the propeller pushes or absorbs the air. The bigger the angle, the more air it can push but this means more power consumption. So for example a ship's propeller can adjust the push of the water just by changing angle of the blades, without changing the rotation speed. The twisted shape of the propellers is because of this reason, as the inner side of the propeller turns at a slower speed than the other edge of the propeller, and in order to get the same push or absorption throughout the blade, the inner side of the propeller has more pitch than the outer edge by transitioning in between. Propellers must be attached in self tightening direction, so that it will not turn loose in time.

Propeller Adapters:

These are attached on to the motors, in order to connect the propeller to the motor. Also called adapter rings.

Propeller Balance:

Propellers are balanced to make sure that the opposite blades have equal weight / moment with respect to center. The heavier side will hang lower, so it can be understood that the propeller is not balanced. Before placing the propeller on the balancer, it must be made sure that the shaft is perfectly horizontal. After balancing, if one side of the propeller is found to be lighter, a thin tape can be added to that side, to make it heavier, until the imbalance is taken care of. An alternative would be to drill small holes at the lighter side and fill with heavy material, or paint the lighter side with compatible paint with enough number of coatings to make that side heavier. If the imbalance is very little, the heavier side can be sanded too. The balancing must continue until the blade can stay stable at any position, not necessarily just horizontal, without any side hanging lower, or tilting smaller than a satisfactory tolerance. Note that the hub of the propeller can be out of balance too. In that case, some glue might be added to the lighter side of the hub.

Photo Credit: DU-BRO RC - www.dubro.com

The propellers coming out of the factory may not always be balanced and it is good practice to balance them before using and even after starting to use, from time to time. Imbalanced propellers may vibrate and harm the mechanical and electronic system of the drone, bearings inside motors over time. An imbalanced propeller will also draw more current because of its vibration, therefore reducing the flight time. The imbalance of propellers will be noticed more at higher turning speeds.

Propeller Guards:
These are attached to the frame, in order to surround the propellers and protect propellers to contact other objects. In many cases, propeller guards are built as part of the frame and not separately attached. In some models, if propeller guards are attached, obstacle sensing system cannot work, as the guards may interfere with the sensors.

Real Time Kinematic (RTK) Technique:
This is a system that is used for positioning the drone with the help of a base station. The base station has a precisely known location and it can be used to communicate with multiple devices at a distance of up to 10-20 kilometers.

Receiver:
See *Transmitter and Receiver*.

Ridge Soaring:
Taking advantage of faster and raising winds, when the wind approaches and goes over a hill or mountain. It is effectively performed by birds, and applying same principles to drone aircraft enables more efficient flight. Also see *Dynamic Soaring, Thermal Soaring*.

Roll:
Tilting the multirotor left or right. This is achieved by turning blades faster on one side and slowing down at the other. See graphic representation under *Pitch*. Also called *Aileron*.

Rudder:
See *Yaw*.

Shell:
Shell is attached to frame as a cover to protect the central core of the drone from water or sun, at the same time improving appearance and aerodynamics. As other parts, it must be made of lightweight but durable material.

Speed:
The speed of the drone in relation to the ground. Generally the speed for hobby grade drones can range anywhere from 15 to 50 miles per hour (roughly 25 to 75 km/h). Airplane drones can have much higher speeds.

Strafing:
Move from left to right repeatedly without changing forward looking direction.

Telemetry:

Telemetry refers to automatic collection of data and transferring of this to a remote location.

Tethered Drones:

Tethered drones are tied to the ground, which offers some advantages. These long lightweight cables can be used for power transmission so one of the main advantages is much longer flight time. Tying of the drone also ensures that the drone doesn't fly away and therefore allows lower skilled operators use their drones without the concern of losing the drone. Tethering can also offer immediate and fast data transmission to the ground from the drone. Having a drone in the air for long period of time, hours or even days, without worrying about the depletion of battery every 10 minutes, offer many advantages and new possibilities that cannot be done with non tethered drones. The main disadvantage of tethering is the risk of entanglement of cable to the drone. Only helicopters or multirotor drones can be tethered, it is not a suitable application for a plane due to high speeds and longer distances.

Thermal Soaring:

Air rises when heated and over warmer areas, it creates lifting force. It is one of the natural soaring principles that can be used, for more efficient flights. Also see *Dynamic Soaring, Ridge Soaring*.

Throttle:

Moving the multirotor up and down, without tilting or turning anything, just by increasing or decreasing the speed of propellers. In other words, it basically controls the level of power delivered to the motors. To visualize easier, throttle is similar to a gas pedal of a car. During lowering of the multirotor drone, doing things too fast may get the drone caught in the downward pushing air that was just created by the propellers and results in drone moving too fast downwards and even crash landings. Trying to throttle up in this situation will cause to vacuum be even greater.

Thrust:

It is the indication of how much weight a motor can lift given a propeller size. Motors that are manufactured for multirotors can have specifications from manufacturers that list thrust values for different propeller sizes. When building a drone, the total weight of drone must be calculated first, then a payload weight is added, if any, finally a safety margin is added, and this will be the total resultant uplift force that is needed. And this value is divided by the number of propellers to find the load per propeller or load per motor. Based on this value, a motor can be chosen or drone dimensions, propeller and motor sizes can be adjusted, until satisfactory values are reached.

Tracking Antenna:

Tracking Antenna autonomously tracks the moving UAV, by directing the antenna to the aircraft, using telemetry data. It is used together with ground control stations and for advanced drone systems and can have over 100 km range. Tracking antenna enables long range data transmission.

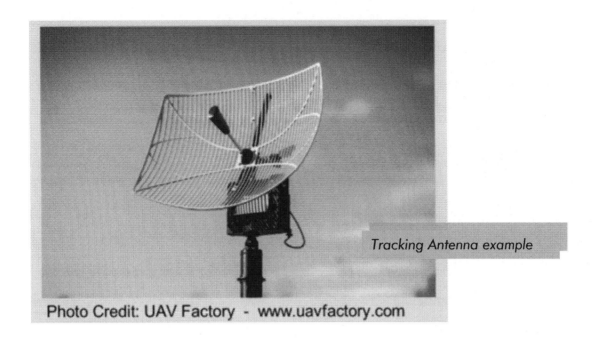

Tracking Antenna example

Photo Credit: UAV Factory - www.uavfactory.com

Transmitter and Receiver:

These are remote control devices used by the drone operators to control the drones from ground. The control of the hobby grade drones are typically achieved through transmitters and receivers using radio waves. Smart phone, tablet or game controller is also possible which use wifi or Bluetooth but they can control only shorter distances. Transmitters come with a receiver which can be first paired with the transmitter, and then, attached to either the motors directly, if manual control will be used, or the input channels of the autopilot, if the autopilot will be used. The typical radio wave length used to operate drones is 2.4 Gigahertz (GHz). Transmitter may come with drone purchase or may need to bought separately. Reaction times and interference free operation is very important when choosing transmitters and receivers, as well as making sure the drone and transmitter are compatible in the first place. On a typical radio transmitter, elevator trim, aileron trim, rudder trim, power indicator, power switch, landing gear switch can be seen. Important note: As the controls use radio waves, you must be aware that radio waves are subject to interference from other sources, and therefore this may cause temporary loss of drone control. So plan your flight accordingly and take your precautions.

Transmitters. Photo Credits: RadioLink - www.radiolink.com.cn

Trim:

On a transmitter there are control rods that can be moved to control *Throttle, Rudder, Aileron, Elevator* moves, which are called as trim.

Visual Positioning:

Using camera to locate the drone to help *GPS* or in the absence of GPS, such as indoors, and rely totally on it for navigation. Cameras can also make a map of drone's surroundings.

Visual Tracking:

Tracking a moving object autonomously.

VTOL:

Abbreviation for vertical takeoff and landing. Helicopters and multirotors can perform this, while airplanes cannot (there are airplanes which can do this but it is exceptionally rare for airplanes).

Waypoint:

These are coordinates, used to identify points. On the surface of earth, waypoints are two dimensional, such as longitude and latitude. In the air, it is three dimensional, with the addition of height dimension. With waypoint navigation, the drone can be navigated through predefined points for a desired path with autopilot. When a waypoint flight path is prepared, it must be remembered that the route needs to be clear of all obstructions, both horizontally and vertically, and the ascend and descend time of drone must also be taken into account for the distance versus time calculations.

Yaw:

Also called *Rudder*. Rotating the multirotor left or right in horizontal. This is achieved by spinning rotors turning in one direction faster and the other direction slower. See graphic representation under the term pitch.

General steps of building a multirotor drone

This section, is especially directed to beginner to intermediate level hobbyists, to build a multirotor drone. Below, main steps are explained.

If you are a beginner, it is strongly recommended that you review the previous section, where terms and concepts about drones were given, if you haven't done already. It is also our recommendation that before making your first drone, buy a small drone and learn how to fly drones. This will help save time and money for the future.

Important Note: In the US you must register your drone if it weighs over 0.55 lbs (250g). This should mean, even if you make your own drone, you must register it, if it is able to fly outdoors. Check with FAA and your state and local authorities. All regulations and safety guidelines still apply for a drone that you built yourself - actually even more, both during and after building.

Build or choose the frame:

For the frame, choose as light material as possible, while maintaining strength. You can also buy a ready built airframe or even 3D print. This is up to you but you must first consider the weight and how many and what size propellers can carry this. No matter what material you choose, proper cutting tools are needed if you build the frame yourself. The lifting capacity is dependent on the motors and the size of propellers. Frame must be able to fit those motors and propellers and the choosing of those are described below. So choosing the frame and motor and propellers go hand in hand as the size and weight of one affect the others. Apart from the size, frame must be strong enough. For example it must withstand the forces of rough landings. Also frames may flex under the force of motors that turns propellers in the opposite direction. Another thing to look is that the frame has as lower center of gravity as possible, to be more stable during turns. For transporting the drone in a small bag later, such as a backpack, or storing your drone, the foldability of the frame is also important, but this is difficult to make properly for a d-i-y drone. If you are building the frame, you can start by attach the arms to the mounting plate that will later hold the flight controller, receiver, power distribution board and the battery. It would be a much stronger hold for the arms, if you use two mounting plates and sandwich the arms between the two plates, rather than just one center plate. The pieces to hold motors at the end of arms should also be attached. After determining the propeller size you can determine the size of your frame, by drawing the 4 propellers on paper or using a CAD software, giving them just enough space between each other so that they will not try to move the same air (note that this distance gets bigger as propellers get bigger), and giving your electronics and camera gimbal (or any payload if it will carry payload) enough space in the middle.

Choose Motors:

First of all it is highly recommended that you choose a motor specifically made for multirotor use. There are thousands of different brand and type combinations and in the beginning it may be confusing. Some motors come with their own propeller adapters and their own custom mounting pieces for easy installation to the frame. The specifications when choosing motors include but may not be limited to the items below:

- KV Rating (rpm/v)
- Compatibility of battery number of cells such as 2S, 3S, 4S etc...
- Weight
- Propeller sizes compatibility
- Maximum surge value (Watts)
- Mounting hole dimensions
- Working current, maximum current, no load current (in amperes)
- Shaft diameter
- Output wires
- Thrust
- Throttle
- Voltage
- Efficiency
- Outside dimensions of motor
- Number of stator poles
- Number of poles (each)
- Internal resistance
- Max LiPo Cell

When choosing motors, the main concern is the lifting capacity. The lifting capacity with different types and sizes of propellers are given. Calculate the total weight of all components plus secondary items like attaching camera with gimbal optionally. You can also attach a parachute, landing gear and a payload, which also add to the weight. As a rule of thumb, a quadcopter with 10 inch propellers, can carry up to 3-4 pounds of payload. This is not considering its own weight. Don't forget that we said quadcopter here. If your drone will have 6 propellers, or even 8, or more, then of course higher payload capacities can be achieved but things get more complicated. Especially if you are a beginner, we suggest that you start with nothing more than a quadcopter. Also add a safety margin on top of calculated total weight and choose the motors per that larger value by dividing the total weight with the number of motors. This safety margin is needed so that your drone can respond to your commands better when maneuvering or in case of emergency. You must do a little research on this and come up with a decision for number and dimension that you will feel comfortable with. Not only the lifting capacity of motors is to be considered but also current requirements too. For a drone, the most amount of current is drawn by motors and other components on the drone draw insignificant current in comparison. So the total of peak current of motors plus a safety margin, must be compatible between motors, ESCs, the battery and it must be something PDB can handle. Note that PDBs cannot handle high currents. In that case breakout cables should be used, which essentially means wiring all the ESC power lines with sufficiently thick wires.

Determine the size of propellers:

This very much goes together with selecting the motors to achieve the desired lifting capacity for a given size of propeller versus given motor capacity. Propellers also must have enough space between each other depending on their size. Too close propellers will try to move the same air which is inefficient. Therefore propeller size affects the size of frame as discussed above.

Choose the flight controller:

Given your needs and what you need to do with your drone, you need to choose a flight controller. At a minimum, things to be considered when choosing are: the possibility to be configured for various multi rotor configurations, has enough sensors for your purpose, and possibly a straightforward setup process. The software should guide you through the processes such as selecting the multirotor type, motor layout and propeller direction, radio and ESC calibration.

Other features of flight controllers may include:

- Assistant software for smart phones
- Failsafe features such as auto return home or auto landing in case of emergencies
- Take off assistance
- Failure protection for motors
- Supporting several inputs and receivers
- Multiple control modes and intelligent switching between modes
- Low voltage warnings
- GPS module for accurate position stabilization, gimbal stabilization
- Intelligent orientation control
- Wireless PID tuning
- Arm/disarm modes for safe motor starting
- Support of PPM S-Bus or general receivers
- On board USB for setup
- Allowing of remote adjustment of gain

The specifications when choosing flight controllers include:

- Weight
- Dimensions
- Microprocessor data
- Sensors (gyroscope, accelerometer, magnetometer, barometer)
- Supported multirotor types and configurations
- Antenna connector
- Supported ESC output with refresh frequency (Hz)
- Recommended transmitter
- Working voltage range including input output voltages
- Interfaces such as input and output data, ports
- Recommended number of battery cells and burst current

- Maximum and normal values of power consumption
- GPS, compass ports
- Operating temperature, hovering accuracy (meters)
- Maximum yaw angular velocity (degrees /s)
- Maximum tilt angle (degrees)
- Ascent and descent rate (meter/sec)
- Built in functions

For more, also see information in the first section of this book, under flight controller.

Choose and calibrate ESCs:

When choosing ESCs, you must choose all of them of the same type, in order to get same response to controls. The main criteria you need to consider while choosing ESC is the current rating, in other words, it must be able to handle the current that the motor requires. Also as usual you must choose the lightest ESC as possible for your need. Note that for peak current calculations for your battery and PDB compatibleness, you must add the peak currents of motors, but not the current ratings of ESCs. The current rating of an ESC is a maximum rated value for that ESC and not the actual value of all motors will draw even at their peak instance. The current rating of ESC must always be higher than the peak current that a motor can draw as a safety margin so it is not the actual value and can be as high as it can, and therefore not relevant in peak current calculation. The specifications when choosing ESCs includes:

- Current rating (A)
- Voltage range (V)
- BEC - if any
- Weight (including cables)
- Motor and discharge plugs and wires
- Size
- Input frequency
- Firmware

After all of the assembly of your drone is complete, the ESCs must be calibrated in order to make sure that they all respond to controls the same way, in other words they all know the high and low points of throttle range. Without knowing the throttle range, for example, you may start the throttle and as you go up one ESC may start and the other may not start yet, and this will of course make flight impossible, therefore this is an important step that must be done. The calibration is done by directly connecting the ESCs to the throttle through the receiver, and introducing the low and high points of the throttle to the ESC. For the further details of how to calibrate the ESCs, there are lot of videos on internet that you can watch, and we will skip the details.

Connect Motors to flight control system:

The flight controller must be setup in order to fully know the components of your drone. This is done via software and then it can be uploaded to your flight controller. The motors are connected to the flight control system via ESC. When connecting ESC to the motor, any one of the three wires coming out of ESC can be connected to the motor. On the other side of ESC, there are wires to connect to power distribution board and to the flight controller, where the ESC draws the power from the battery, and with the signals that are sent from the flight controller, adjusts the amount of current that goes to the motor. The black wires from ESC should be connected to the negative pins and red wires to positive pins of power distribution board, just like the black and red wires from the battery. Power cables between motor and controllers must be as short as possible. Also see the photos under the ESC description in the first section of the book.

Connect receiver to flight controller:

The flight functions such as pitch, yaw, throttle must be connected to each other by pairing appropriate pins. The receiver and the flight controller are placed near each other on the top mounting board of the frame and attached to it. Also, when placing receiver onto frame, it must be kept as far as possible from the power system (motor and battery) to prevent interference. At a minimum it should be kept at a distance of several centimeters from motors or servos.

Choose battery and connect to power distribution board:

For a normal hobby drone, you may choose a 3 cell battery. If each cell generates 3.7 V, it will give a total of 11.1 V. This voltage can later be regulated with the help of voltage regulators or BECs, to different parts of the drone. You must choose a battery that has the same number of connectors as power distribution board. If not, an adapter can be used but this will add to the weight. The power distribution board delivers power to all motors, separately, which all have their own ESCs. The board must be able to handle the total of all peak charges that ESCs draw to transmit to motors. If not, cables can be used, for handling high currents, instead of a PDB. When choosing a battery, you must remember that larger batteries have higher capacities but they are also heavier. Your flying style, weather conditions, weight and center of gravity of the drone, rpm of the motors, and the size and pitch of the propellers are factors that affect battery life. Safety note: when dealing with the battery be careful for not to cause any short circuits and follow specific instructions for battery.

The specifications when choosing battery includes:

- Minimum capacity (mAh)
- Voltage
- Number of cells
- Charge and discharge plug
- Constant and peak discharge
- Weight
- Dimensions

Connect flight controller to frame:

In order to reduce the vibration that the flight controller gets from the frame, vibration isolators such as rubber can be added. The flight controller can be attached to the frame by either sticking it to the frame or using screws.

Attach Battery to frame:

The battery, being the heavy component, greatly affects the center of gravity of the frame so its location can be adjusted as needed. The best location for the battery is the center of the drone so that the motors can share the weight equally but of course this depends on the location of other components too.

Attach motors to frame:

The motors are first attached to a mounting plate that will hold the motors which would fit to the end of the arms. After attaching these plates to the motors, they should be together connected to the end of each arm.

An assembled drone, with autopilot, ESCs, motors and propeller attached to the frame.
Photo Credit: GetFPV - www.getfpv.com

Attach payload:

This is an optional item, in cases where you want to attach a payload to the frame. You must make sure that the payload will not shift during flight and distort the center of gravity of the drone. The attachment points must be firm in order to avoid accidental drops or slips during the flight. Remember that when the drone tilts the most and the payload is the maximum possible, (the worst case) this will create a big turning effect, (moment = force x distance) at attachment points when your drone rotates, so do not think that those connection points will only carry a vertical load and make your connections strong enough. Plus there are additional inertial forces due to acceleration and turning of the drone.

Attach propellers to motors:

Propellers are attached to motors by adapter rings, which are also called propeller adapters. The adapter ring comes with the propeller, or you buy it separately but make sure that the inside diameter of the adapter ring fits the motor shaft. The propellers need to be balanced before attaching to motors. Unbalanced propellers will cause vibration for the drone which is bad for many reasons including damaging the motors, loosening of attachment of all components to frame, and cause photos and videos to be of poor quality. Balancing propeller with the motor at the same time can also be done if necessary, using more advanced techniques. For a quadcopter, prop 1 and 3 should turn in the same direction, say, clockwise (CW), and props 2 and 4 should turn in the opposite counter clockwise (CCW) direction. Choosing the right propeller can make a difference in the drone's ease of control and even battery performance so you can try different pitch propellers even for the size you determine, or you may end up with a slightly different size propeller than you considered before with a different pitch or even different number of blades. When buying propellers and motors it is a good idea to order extra pieces. There is a chance that you may break propellers or one of the motors may not work in harmony with the others.

Attach antenna to frame:

Antenna must be placed as far as possible from cables, carbon fiber and metal parts. It must also be electrically isolated from the frame.

Attach the casing to the frame:

Finally the casing should be added as a cover in order to have better aerodynamics as well as protecting the flight controller. Another purpose of casing (shell) is to have an aesthetic appearance. Remember that there are different ways that you can improve the aesthetics of your drone, such as painting the frame and casings, attach led lights, which consume very little power and they weigh very little.

Choose Transmitter:

The specifications when choosing the transmitter includes:

- Number of channels
- Battery compartment
- Display screen
- Memory
- Encoder type

Also make sure that you choose the right transmitter depending on whether you are right or left handed. Most transmitters have throttle on the left.

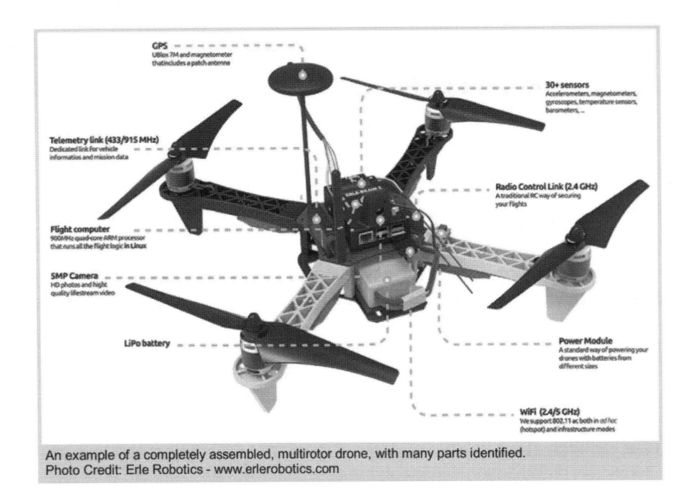

GPS
UBlox 7M and magnetometer
that includes a patch antenna

30+ sensors
Accelerometers, magnetometers,
gyroscopes, temperature sensors,
barometers, ...

Telemetry link (433/915 MHz)
Dedicated link for vehicle
information and mission data

Radio Control Link (2.4 GHz)
A traditional RC way of securing
your flights

Flight computer
900MHz quad-core ARM processor
that runs all the flight logic in Linux

5MP Camera
HD photos and hight
quality lifestream video

LiPo battery

Power Module
A standard way of powering your
drones with batteries from
different sizes

WiFi (2.4/5 GHz)
We support 802.11 ac, both in ad hoc
(hotspot) and infrastructure modes

An example of a completely assembled, multirotor drone, with many parts identified.
Photo Credit: Erle Robotics - www.erlerobotics.com

A multirotor drone, showing how parts fit together.
Photo Credit: Erle Robotics - www.erlerobotics.com

Things to consider when buying drones:

Apart from the cost, and there are many important criteria, and points to pay attention when buying drones. These include:

- Hovering feature
- GPS navigation functionality
- Autonomous flight from point A to B
- Automatic takeoff and landing, which can help protect drone from damage, in the event that a relatively inexperienced pilot is operating the drone, or just for convenience
- Operational time, speed and range, with and without payload
- Ability to return to base in the event of an error or user command
- Position holding ability, even with strong winds to keep the drone stable at a given location. One further step to this is the dynamic position holding, which allows a operator to keep the drone on a straight line, between waypoints, even with strong winds
- Following a moving person or object autonomously at a desired altitude, distance, bearing and angle
- Orbiting ability around a given target point
- Targeting a certain location and altitude
- Avoiding obstacles in the air autonomously, indoors or outdoors, by also using cameras and ultrasonic sensors
- Ability to save a given position in the air and flying back to it when desired
- Focus camera gimbal at a point of interest, while the operator directs the drone
- Ability to perform waypoint navigation, where the drone can travel along specified waypoints for a desired travel path. These points can be preprogrammed using a laptop or tablet
- Ability to trigger the camera shutter at the previously specified locations during the flight, in relation to waypoints
- Taking panoramic photos of a target by directing the flight controls and camera gimbal accordingly
- If you are buying from a specialty store be careful about the return handling. Make sure that the company will not direct you back to manufacturer country.
- Whether you are buying or building, it is always a good idea to visit one of the forum sites on multirotors or drones and ask advice or at least confirm what you are thinking before you spend money. Even if you are sure, you may be surprised with the answers you get from many experienced people.
- Parts that have higher possibility to break or not function precisely, such as motors or propellers might be ordered in excess to make sure you have spare.

Aerial Photography and Filming Overview

We also want to touch up basics on aerial photo taking and video recording, as most of the time drones are used in filming and photography.

Discussing cameras, photography and recording videos is a whole other subject by itself and out of scope of this book. It is only presented here as an overview, particularly in relation to filming with drones and we will leave it at that.

Some of the important items to be considered when taking photos or videos with drone:

- The main thing to consider when choosing a camera is to know whether you will mainly take photos or videos and where will you take them.

- There are different types of cameras for different purposes. The camera to choose for taking wedding videos should be different than a camera needed to take landscape photos, monitoring a ranch or a fast action snowboard video.

- The camera you use must be lightweight and it should be able to stream videos smoothly.

- Frame rate, watertightness, megapixels, dimensions and of course cost are all important factors to consider when choosing.

- The shaking caused by the moving drone must be countered in order to get still images or videos. A gimbal is a good solution as it stabilizes the camera. Also, propellers must be balanced before the flight, in order to reduce shaking caused by unbalanced propellers. For more information, see propeller balancing term.

- The light source, sun's position must be considered, just like when you are taking a photo on the ground yourself.

- Taking photos or video during sunrise or sunset hours can create very attractive looking results, as the sun illuminates buildings and natural features differently and the shadows are more defined, adding to the attractiveness of images or videos.

- The wind speed increases at higher altitudes, even if it may not be windy at the ground level, since the friction of air and land cause the wind speed to drop at the ground level and rise gradually. Wind causes shaking of the drone, so this must be kept in mind.

- The object of interest must be followed by the camera with proper angle, depending on the position of the drone.

- For taking photo or video at night, in addition to headlights, to illuminate the target, colored LED lights should be placed on the drone, so that the operator can see the drone and its orientation.

- The photo or video shoot must be planned ahead of time, such as clear line of sight that must be maintained always, the timing of video or photo taking, which must be shorter than the flight time, by taking takeoff and landing times into account.

- Rainy and windy weather should be avoided if possible.

- The photo of the same target can be taken with different camera settings, so that the optimal image can be chosen later. This is called bracketing.

- The photo or video of the target must be captured from different angles, and not from just one angle, to obtain best results.

- Using a gimbal makes the photo or video quality better, as it stabilizes the camera, but consider the weight of it versus drone's payload capacity when buying a gimbal.

- The sun's position is important not just for the quality of the video but it may also cause the propellers to cast shadow on to the camera. As a general rule, filming towards the sun should be avoided.

- The landing gear must not get in the way of the camera view so it must be chosen accordingly.

- Some drones come with non interchangeable cameras, so you must be sure that the camera will serve your purpose, if you are buying such a drone.

- FPV, first person viewing capabilities of a drone system is also an important factor when choosing a drone for filming or photography.

- The speed of the drone should be low, as well as acceleration, deceleration and turning of the drone should also be slow, in order to avoid shaking and get better images or video.

- Although being close to target is important, it is also a good practice to increase altitude and capturing more area in the photo or video.

- In addition to moving the drone, the gimbal can also be moved at the same time, for more attractive videos.

- The images should be taken in RAW format, which means it has the least minimum processed data from sensors, so that they can be optimally processed later, such as the colors and the exposure of images. A RAW image is an image format that has all the information in it to create an image but it is not yet ready to be printed and needs processing. This gives the ability to optimally process images later. RAW images can be considered as digital negatives.

- When capturing photos or videos with a drone, you must make sure that you are not trespassing into private property or violating privacy of others.

- Try to move the drone through two, such as downwards and forward at the same time, for more professional looking videos. Orbiting around the target and strafing of drone also creates attractive videos.

Drone Safety Overview

Warning: This section is not an all inclusive list, and it is written here for the purpose of giving examples and raising awareness about safety. Check your particular situation with drone manufacturers, neighbors, other experienced drone users, professional drone consultants, and your local, state and federal authorities as applicable. Flying a drone must not be viewed as operating a toy, and it requires careful operation and planning by a knowledgeable and skilled operator.

First of all, if you are in the USA, you should be aware of the laws by FAA, Federal Aviation Administration. https://www.faa.gov/. FAA says that you become part of the US Aviation system, when you fly a drone in nation's airspace. The list below, which is shown on FAA website, gives some basic guidelines:

- Fly below 400 feet and remain clear of surrounding obstacles

- Keep the aircraft within visual line of sight at all times

- Remain well clear of and do not interfere with manned aircraft operations

- Don't fly within 5 miles of an airport unless you contact the airport and control tower before

- Don't fly near people or stadiums

- Don't fly an aircraft that weighs more than 55 lbs

- Don't be careless or reckless with your unmanned aircraft – you could be fined for endangering people or other aircraft

- Drone operators must be at least 13 years old to obtain a certificate

For all other information and more details, permissions, certifications, checklists, reports, we suggest that you carefully check FAA website. There are also many firms that give guidance and provide consultancy services, in order to help on these issues.

Drones also should not be used near wildfires, because they can be dangerous for firefighters, as explained in USDA website.

Various other safety guidelines include but not limited to:

- Drones should be operated at a safe distance from vehicles or other people.

- Batteries must be removed after operation.

- Moving parts must be kept clean and dry.

- Drones and drone parts must be kept out of reach of children.

- Moving parts must not be touched.

- Do not operate a drone under influence of alcohol or drugs.

- Wait for a strong GPS signal before takeoff, in order to take advantage of safety features such of returning to base location and position stabilization.

- Drones with damaged parts or wiring must not be operated.

- Do not expose any equipment or parts to water, unless it is specifically designed for it.

- Turn the transmitter on first, before turning on the drone, and after flight, turn off the transmitter last, after turning off the drone.

- Check battery manufacturer's manual and abide by all the safety precautions about using or charging batteries. Remember that improper usage or charging of batteries may not only result in poor performance or falling of your drone, but also can cause fires.

- Make sure you do not have distractions when flying.

- Understand the safety functions of the flight mode that you are using.

- Drones should not be used in snowy, rainy, foggy or too windy weather.

- Prefer open plain areas for flying far from people.

- Make sure that there are no water sand oil or any other foreign objects in the drone.

- Do not talk on telephone when flying a drone.

- Stand at a safe distance from the drone during takeoff.

- Chipped propellers must not be used.

- Do not try to modify motor structure.

- You should use parts that are recommended by your drone manufacturer, and if in doubt, you must contact your manufacturer about the compatibility of a certain part.

- If you think that your drone is definitely about to crash, turn the throttle to zero.

- Flying near broadcasting towers, high voltage lines is not safe, as these affect magnetic and radio waves.

- Remove all batteries from the drone, before doing any work on it. If for some reason you must keep the battery there, then remove all the propellers before doing any work. The propellers of a drone can damage your hand badly.

- Initiate timer when starting flight, if it is not automatic.

- Do not trespass private property.

- The drone must be landed as soon as possible, if it starts to drift excessively. Chances are the compass is malfunctioning.

- Start flying with full batteries. Even if the indicator may show partially full battery, it may not be accurate and you may have less than expected flight time, when you start with a partially full battery.

- The higher you fly, the more careful you must be. It also gets windier at higher altitudes.

- When recording with camera, respect privacy of others.

- Read and understand all safety instructions for your drone from the user manual.

Drone operators should also log their flights into a flight log to keep records of the flight such as date time location etc... Check the official requirements about this. There are even online services that help doing this online.

The rules and regulations on drones are still evolving as this is a new area. As of end of 2016, FAA requires all drones which weigh over 0.55 pound in the United States to be registered. The drones operated indoors only however, do not need registration. You can also check the FAQ section on FAA website at: https://www.faa.gov/uas/registration/faqs/

Future of Drones

In the future drones will have increased power efficiency, battery strength, reduced noise, increased operation time, increased speed, altitude and range, increased autonomous abilities. These will also increase areas of use and carrying capacity. Many future possibilities greatly depend on battery power.

Size of drones will decrease which means they will have less kinetic energy on impact in case of a fall and therefore it may ease regulations.

Manual piloting of a drone will be history, unless specifically desired by the user.

Delivery drones may be common place. Drones will be able to carry more weight, and drone parachutes may be a mandatory feature. There may be drone charging poles at every certain distance, where, drones automatically attach and recharge. In the future the volume of goods delivery will increase, even if we will have 3D printers at every house or office and producing more and more variety of items. Drones will take over some of this increased burden from cars or trucks. Many people will be needed create, run and to manage every aspect of drone business. This will be one of the major effects of drones on future economy as well.

We may have our own drones, similar to having cars, to pickup any items we need, when we want, not only for shopping but also anything else too. With increases in flight time, we may also see our personal drones following us to constantly from a close distance to record our activities, such as when we go mountain climbing, skiing, sailing...

Data gathering about our surroundings will greatly improve, even in real time.

Construction: Attachment of arms, or tools, which will open a whole new world of possibilities, such as in construction. There are countless tasks in construction, where materials must be elevated or workers must work above ground level, or, as in most cases, both of these taking place at the same time. Drones can carry things up and down, and skilled craftsmen on the ground can direct the drone arms through FPV, or some tasks may take place autonomously with the improvements in robotics. Going even further to the future, the autonomous abilities of robots will improve to the point that they may not need humans, to direct its arms when doing a certain construction installation. So all they may need is a construction plan or sketch uploaded to them, and then they will do the work autonomously, also by communicating to each other. Also remember that one of the biggest challenges in robotics is to make a robot move itself on land in an undefined environment. With the use of drones that fly, this challenge is eliminated altogether. All this will increase construction speeds and reduce costs. Drones are already being used for visual inspection of elevated surfaces where normally a crane or scaffolding would have been required to access.

Law enforcement applications will increase due to increased battery power and autonomous abilities, such as tracking of suspects, surveillance, remote presence.

Military Applications:

- Air minefield, or swarm, such as a network of small drones in big numbers swarming around enemy airplanes, helicopters or rockets to force them to change direction or interfere with them. A similar application of this was already tested on the sea by US Navy in 2014 which involved small autonomous ships in big numbers swarming enemy ships. Keep in mind that the drones of the future will be able to stay in the air much longer time, for much cheaper price. A big network of totally autonomous small drones, spacing themselves at a certain distance from each other and constantly communicating location of each other by using swarm robotics concepts and location and projected path of enemy planes can provide an air defense mechanism, by communicating the projected path to drones ahead. With small guns or bombs attached, the effectiveness and coverage radius will increase, or they can even use electromagnetic waves to jam the electronic devices of the enemy aircraft that passes by. Air swarm concepts can also be used for attack type of force too, of course.

- Manned warplanes will be eliminated greatly or entirely, and be replaced by drone fighter jets. The battle of two countries air forces may only mean their drones are fighting. An unmanned plane will also not have the acceleration limitations as a manned plane, since the human body can withstand up to a certain amount of acceleration.

- Drones may operate very close to land surface, with light weapons or bombs attached, to act as land soldiers, for more precise operations than today's armed drones, which can only shoot from long distance. The advantage of these against a walking unit would be that they will be much faster than a walking unit, plus they will not need to tackle the challenges of walking on irregular surfaces. They will be very close to surface, but still take advantages of flying. The movement and weapon operation of drones can be autonomous or remote controlled as necessary.

In 2050, with all the technological improvements achieved and regulations long settled, we may see many drones everywhere, capable of doing much more variety tasks with more abilities, flying much longer time and carrying heavier loads.

Unmanned Ground Vehicles (UGV)

Although most of us think that an unmanned ground vehicle is basically an autonomous car or truck, there are other types too. Below we discuss each of these categories.

Automated Guided Vehicles (AGV):

AGVs navigate by following certain markers on the floor such as wires or guide tapes, or find its way with lasers or vision. Most of the time they are used in industrial applications such as moving around materials in warehouses. Automated guided vehicles (AGVs) are also known as LGV (laser guided vehicle), SGV (self guided vehicle) or VGV (vision guided vehicle).

Autonomous Cars (Driverless Cars):

Also called driverless cars or self driving cars, these autonomous vehicles appear to be coming into our daily lives soon, which will have a huge impact on how we transport ourselves and in turn a major effect on our way of life. Indeed, in the Federal Automated Vehicles Policy, issued by US Department of Transportation, it states:

"The development of advanced automated vehicle safety technologies, including fully self-driving cars, may prove to be the greatest personal transportation revolution since the popularization of the personal automobile nearly a century ago."

Therefore, it is clear that autonomous driving is one of the most important robotics related technologies that we can discuss.

Autonomous cars need to do basically two things to find their way and drive: Create the complete map of its surrounding area including the objects and the travel path defined in that area, and determine its relative position and what it is doing with respect to that defined map – here defined means that the car "knows" the meaning of the objects in that map. Of course this map and the relative position of the car versus that map is dynamic and being continuously updated. In order to do all these, an autonomous car uses equipment such as:

Radar sensors: Radar sensors are mainly used to detect various obstacles

Cameras: Currently used mostly for distinguishing the lanes and backup assistance, but as the image processing software gets more developed, the importance of cameras on board are increasing.

Image-processing software which can detect traffic signs and lights, lane stripes, and other objects.

GPS Units: Global positioning system is used for determining a car's location by getting input from satellites.

Accelerometer, inertial navigation systems: Help with navigation of the car when the signal received from GPS devices are poor.

Ultrasound sensor: Currently ultrasound sensors are mainly used for detecting obstacles in front and back of the car while manually or automatically parking the car.

Wheel sensor: Also used in stability and anti lock braking systems, another use of the wheel sensors is to keep track of vehicle's location when the GPS systems are temporarily unavailable due to poor signals.

Laser Range Finder (Lidar): Lasers that spin in order to constantly take horizontal distance measurements. The information taken from these measurements are combined with the information coming from cameras and the radar in order to create a detailed map of surrounding. With this sensor taking so many measurements of the immediate surroundings of the car, a detailed 3D map can be produced.

Also see *Autonomous Navigation, Robotic Mapping, SLAM*, under *Robotic Terms and Concepts* section.

Benefits of Driverless Cars and Future Predictions:

Some of the benefits of self driving cars may seem obvious to anyone who has even the slightest idea about technological developments, but let's cover it with some facts, and in more detail to get a better picture, including predictions for the future.

Reduced Accidents: Each year, an estimated number of 1.3 Million people die in traffic accidents, which is the 10th leading cause of all deaths overall, and 50 million more suffer injuries, according to a Wall Street Journal report based on World Health Organization data. Widespread use of autonomous cars will reduce this number, because the leading cause of all traffic accidents is human error. Of course we are assuming that by the time the unmanned cars are allowed for general use, a lot of safety measures will be in place in addition to more improved technology than today, so that the "human error" will not just be replaced by the "machine error". Even if there are rare machine errors and they cause deaths or injuries, the total will be far less in numbers, in comparison to what we are seeing today. Therefore, we suspect that when a machine error causes accident, this will not be a cause to disallow the whole autonomous driving concept, because as a society we will see that we are still benefiting in the end, if we consider the greatly reduced death numbers. This is similar to flying by airplanes, where we suffer from unfortunate accidents from time to time, which are usually the fault of the equipment, rather than the highly trained pilots, but regardless, we continue to fly.

Statistically flying is the safest way to transport, if you think in terms of the number of passengers carried versus the losses. When autonomous cars arrive on our streets, even this might change. To be able to transport ourselves safely is becoming always easier to achieve for us, as we advance our technology. Autonomous driving is just a natural and logical step to that, just as everything becoming more and more automated, have more information in it than effort.

Traffic reduction: Machines are very precise. They are incredibly fast in reacting too. Think about a highway with heavy traffic where cars stop and go… Each time a car moves, some seconds are lost between two cars. Multiply this by the number of all the cars on the highway. You reach to a very large number in terms of delays. Plus humans need more safety gap in between due to slower reaction time. With robotic cars, this inefficient process will be history. They will be able to react instantly to the moving traffic ahead with closer distances to each other, and this will create a much more efficient and continuous flow of traffic, which will increase highway capacities, even in packed situations. It will essentially create a "train of cars" on a highway. Also remember that it is not only the reaction time or shorter distance of the individual cars in question here. By swarm robotics concepts, these cars will also be able to communicate between themselves, and even with the surroundings, thanks to chips becoming cheaper than water and smaller than dust and they will very easily be placed (may be even by spraying at some point) on every physical thing we can think of, which leads to further improvement of the communication process, increasing the safety and efficiency of driving. By the way, the chips making everything around us smarter and our physical world as one large living organism made of information like an internet of physical things is a whole new subject by itself which is not too far from robotics but out of this book's scope.

Higher safe speeds: As the reaction speed and safety of autonomous cars are far greater than humans, the speed limits may be increased.

More space and easier parking: The parking process will be much easier both in terms of space and time. Someone who needed to find a parking place before, will just be dropped off wherever he wants and his car will park itself at a location where parking space is abundant. This will save the passenger's time and will also help solve parking space problems as the car may park far away and come back when it is needed again.

Traffic police: There will be a dramatically reduced need for traffic police, if at all. This job might be done by a system setup in the intersection of a crossroad, where a set of cameras and sensors will control the intersection and communicate with the cars wirelessly.

Insurance: Car insurance premiums will decrease. The main cause of higher premiums is accidents and reduction in this number will make premiums cheaper.

Time saving: Needless to say, the most obvious benefit about autonomous cars that first comes to mind is that instead of spending time by paying attention to the road, you will now be able to do something more productive in your car, such as reading the latest celebrity news to see who is dating who.

Less number of cars and lower costs: Overall, there will be reduced number of cars needed and the average cost of transportation by car will decrease. One reason is the elimination of a redundant passenger, such as the driver, which will not be needed. This will in turn increase the carrying capacity of the cars, which means less cars will be needed, no taxi or bus or truck drivers needing to make money, and it will also save on fuel overall, as the weight of one passenger will go away and less cars operate on the road as the driver can be replaced by the passenger. Overall, this will not mean less jobs though, as much more people will be needed to work on more information based jobs.

Another contributing factor to the decrease in number of cars is that the people will be able to lend, rent and borrow cars easier, as the cars can just drive where they are needed. At present, most of the time our cars just wait for us uselessly, occupying parking space. But imagine them being able to drive and carry others instead of just waiting for us. This will not happen always, still you might want to keep your car just for yourself but it will happen anyway and we are talking about average here. It means the operational time of cars on average will increase, which in turn means, the same total amount of transportation we need, will be achieved by less number of cars. Today even if you wanted to lend your car to someone, he or she would need to come to your physical location to get your car and the keys. This will actually make it kind of redundant and very inconvenient to get your car because in order to get to where you are, they would need to use another car or at least some sort of transportation. Even car ownership may decline. Car renting, borrowing and taxi concepts will be transformed this way. You may not even have to be near your car to start your car. Just enter your credentials by a phone app or on the internet, and it will start your car for you through its internet connection and you tell your car where to go and when to come back. There may even be internet sites or phone apps arranging all these instantly between people who want to lend or borrow or companies who run these cars as autonomous taxi. Or you just go to the street the old way and pull a cab – without a driver, which basically operates for much longer hours than a regular cab as it has no driver to wait for when he sleeps or eats, which means less taxi on the road also. You will have daily or monthly tickets for car usage such as metrocard or train or bus pass of today, where the local municipality or a local taxi or rental car company can provide for you. Just swipe your metrocard or "autocard" when the car arrives. For example, Avis already invested in the concept, buying Zipcar for nearly $500 M as of January 2013, which provides car sharing services as described in here, without the automonous cars - yet - of course. Or Uber is already starting to test autonomous cars on the road. When the technology arrives, the infrastructure is already in place. The details of the whole system will naturally smooth itself out in time.

When talking about less number of cars, we should not forget the market forces too however, in other words, if the number of cars drops too much, then the price for car will increase to the point that owning one will be cheaper for some. So in the end it will balance itself. We must not forget that the cost of manufacturing physical things always decreases, which will be true for making cars too. But just because it will be cheaper, doesn't mean we will have unnecessary number of cars on the roads than we need totally as nothing is still free and also think about the cost of registration, maintenance and insurance which will still be in place.

Improved transportation of goods: Autonomous cars can even be sent to do the tasks that will not need to carry passengers at all, but just goods. Someone will be able to order goods online or by phone and then just send his car to pick it up, if she doesn't want to wait for delivery or pay for shipping, where the seller will just load the goods into the car for pickup. Or the seller might do the delivery, like today, but as there are no drivers needed, the shipping cost will decrease. This can even be combined with drones, such as one truck carrying goods to a general area and the drones it carry can do the deliveries from air and return back to the truck. So the retail and shipping industries will be impacted also. In many cases when we go shopping, we drive to a retail shop, just to load the goods into our car and bring it home, unless we want to see or do something inside. Though when we go there, often we do or buy something else than we originally planned for. Eliminating all this and just sending your car to pickup things will therefore have effects on the retail industry. It will also mean more free time for you at home.

One person cars: As discussed, the taxi driver is eliminated by autonomous cars, and for taxi, single person cars will also be suitable. Even for buying, single person cars may be interesting for many people. These cars, due to smaller size, can also improve traffic and parking efficiency, in addition to the overall reduced number of cars.

Impacts on economy: Fear not, autonomous cars do not mean we are loosing jobs to robots. As always happened since industrial revolution, each automation creates higher quality and more information based jobs even if it eliminates some old professions. Just like the industrial revolution replaced almost all people working in the farms with machines, who started doing something else, other professions which have been created by the new technologies themselves. For instance in this case, we will not have taxi drivers anymore, but more people will be needed to create and manage the software, technology and the process.

Less cars mean less auto mechanics. Robot cars make less accidents too, and they drive less abusively and in an optimum way, which means less repair jobs per car also, except the regular maintenance jobs needed. Add to this fact that our ability of manufacturing of goods is always perfecting itself and we are able to make things more durable with better systems and materials needing less replacement. All of these combined, a huge decrease in the number of auto mechanic shops can be expected by the time autonomous cars become mainstream and that trend will continue. Less auto mechanic shops mean a decrease in the automotive parts and accessories economy also. This seems like a big deal and will have a huge impact on the economy like a domino effect, as automotive industry is one of the locomotive industries of the economy, affecting so many other industries. But again, all these lost jobs and economy, will be compensated by the new professions created by the new technology plus the efficiency added to our lives and increasing our production.

Environmental benefits: Less number of cars also mean less fuel costs, less material costs, less traffic, which is obviously a positive impact for the environment.

In September of 2016, US Department of Transportation issued guidelines for automated driving, which is called the "Federal Automated Vehicles Policy". It is a 116 page document that can be accessed here: https://www.transportation.gov/AV. We would like to highlight some of the points that are mentioned in this document in regards to future benefits as below:

- 35092 people died on US roadways in 2015, and 94 % of these crashes can be tied to a human choice or error.
- A human driver can repeat the same mistakes, but a Highly Automated Vehicle (HAV), can learn from previous data.
- System performance can be improved by using Vehicle to Vehicle (V2V) and Vehicle to Infrastructure (V2I) Sensor technologies.
- Disabled or elderly people as well as people who cannot afford to buy a car will have better access to cars
- Infrastructure capacity will be improved without pouring a single truck load of new concrete, due to the possibility of better organizing the existing infrastructure
- Energy Savings and Reduced Air Pollution will be realized

So the conclusion is: fewer cars, lower costs, more efficiency, increased highway capacities, increased speed limits, better parking, less idle time of cars, as this whole thing simply becomes another automated and redundant process in our lives, which used to be inefficient and tiresome in the past (today). Just like excavating a foundation pit in the past by hand, versus using the big excavators today (even construction equipment will be autonomous, but that will come definitely after autonomous cars, as in that case the AI must replace a skilled operator who used to perform much more complicated tasks in addition to what a regular car driver does). Overall, if you think about all these benefits counted above, you will see that we are looking at a much more improved picture of transportation with a lot of positive impacts on the society. In the end, this will just be viewed as any other simple automation that happened in our lives, like many other things that we take for granted today but is automated already, such as robots making cars in factories. Finally, in 2050, at a time when manual driving will for so long be history, our children will look at the old movies of today, and our manual driving of cars will seem to them the same way we perceive horse carriages today.

More on autonomous cars:

In 2013, the US Department of Transportation's National Highway Traffic Safety Administration (NHTSA) defined five different levels of autonomous driving:

- Level 0 - No automation: All primary vehicle controls (brake, steering, throttle, and motive power) at all times are controlled by the driver only, and driver is solely responsible for monitoring the roadway and for safe operation of all vehicle controls. Vehicles that have certain driver support/convenience systems but do not have control authority over steering, braking, or throttle would still be considered "level 0" vehicles.

- Level 1 – Function-specific automation: Automation at this level involves one or more specific control functions and even if multiple functions are automated, they operate independently from each other, where in no instance they replace assume the full control of the vehicle but merely assistive in nature to the driver. The driver still has the overall responsibility from safe operation of vehicle and has the main control, but sometimes can choose to delegate a primary control manually as in adaptive cruise control or lane keeping, or automatically in emergency and crash imminent situations such as electronic stability control or pre-charged brakes, where the vehicle automatically assists with braking to enable the driver to regain control of the vehicle or stop quicker than by acting alone. In this level of automation, the driver <u>never</u> has hands off the wheel <u>and</u> foot off the pedals at the same time.

- Level 2 - Combined function automation: This level involves automation of at least two primary control functions designed to work in unison to relieve the driver of control of those functions for shared authority in limited driving conditions. An example of combined functions Level 2 system is adaptive cruise control in combination with lane centering. In this mode, the driver must still be available at all times even during the instances of autonomous driving, and still has the sole responsibility of monitoring the road and safe operation of vehicle but he or she <u>can</u> take hands off the wheel <u>and</u> foot off the pedal at the same time, when automated driving is in operation.

- Level 3 - Limited Self-Driving automation: This level of automation enables the driver to cede full control of all safety-critical functions under certain traffic or environmental conditions and in those conditions to rely heavily on the vehicle to monitor for changes to determine if the driver must regain control of the vehicle. The driver is still expected to be available for occasional control, but with sufficiently comfortable transition time.

- Level 4 - Full Self-Driving Automation: The vehicle is responsible for both monitoring road conditions and safe operation of the vehicle at all times. A driver may or may not be present but never has the responsibility to operate the vehicle and may only be required to provide navigation information input for the destination point.

Note that in some places it is possible to see these levels as 1 to 5, instead of 0 to 4, as written here.

In December 2016, a real life video taken from the camera of a Tesla car showed that, the car warned its driver about an imminent accident, not even for the car itself, but *between the two cars traveling in front of the car*, by sensing *the front of the other car in front*. A human driver would hardly have seen this coming. The car warned its driver by a beeping sound that the two vehicles ahead was about to collide with each other, and immediately applied the brakes, before the driver himself could, and yes, the two vehicles ahead really did collide after about half a second. Again, the two other cars were traveling in front of the car in question on a freeway, not coming from opposite direction or sideways, and it was hard for a human driver to detect this coming plus act on it as fast as the computer, so the car prevented an accident that a human driver might not have or do it with difficulty. (Of course if a human driver follows the vehicle in the front with enough distance versus speed, no matter what happens in front will not cause accident as the driver will have sufficient reaction time, which is the rule followed by drivers since 100 years, but we are talking about an example here). This video went popular on social media because it showed that an autonomous car not only can drive as good as a human, but sometimes even better, where it can detect threats a human may not or may hardly do so, and act on it faster. This was another clear sign that the autonomous cars will be mainstream soon, as they are already starting not only to match, but outperform humans. One can imagine that in the future when the majority or all of the cars are autonomous, in addition to the safety level described here, the cars would also be talking to each other wirelessly, making the process even safer. Actually it seems to us that shortly after autonomous cars will be mainstream, wireless communication of cars in close proximity with each other will be a necessity, a mandatory feature required by law, not only for improved safety, but also improved driving performance and more efficiently using roads and fuel.

As of 2017, there are still legal and safety challenges about making autonomous cars mainstream, however it seems that the required technology is completely here (although still somewhat expensive to be mainstream, but the cost will surely go down in the coming years) and many firms have long started testing their driverless cars on the streets, whose autonomous cars have already driven millions of miles in total. In March 2017, BMW announced that they will be able to make level 4 full autonomous cars by 2021.

The laws surrounding the driverless cars vary from state to state, for example as of January 2017, while in California the companies need special permit to operate an autonomous vehicle, and about 20 companies including Mercedes-Benz, Google and Tesla have such permits, in Florida an autonomous car is in no way treated differently than a regular car and there is no state or local permit needed (please still check with the officials, if you intend to operate one).

Note that, no matter how much we perfect the autonomous technology, and even when cars ability to drive far exceed that of humans, there will still be a need of human input time to time. For example, when you arrive at your destination, you may want to stop your car at a certain exact location, for any reason, such as parking below a tree shade, or directing the windshield at a certain angle to watch a nice view, or at a certain distance from a garage door any other object, or while on the road you may want to make an extra stop in front of a shop, parking at a crowded location where in the last moment you see another car was waiting and you give your way to him, you decide to stop near a stranded vehicle, and many small details like this, which would not be the decision of your car. These detailed decisions must still be made by humans, and there must be a way to make the car to do these, especially when the manual driving will be long history (as human driving ability will be inferior in comparison to the car itself plus human driving will remove the benefit of the cars wirelessly communicating between each other and therefore may not even be allowed anymore) so still there will always be a way to control a car by human, at least in certain allowed circumstances, no matter how far we go into the future.

For cars, as the technology advances, not only autonomous driving capabilities, but also other features are advancing. Car companies are trying to keep up the pace with many new features that new technologies make possible, such as the ones that are readily available on smart phones, but the difference here is that these systems must be able to respond to the user quicker, where he or she is traveling on a highway and the controls must be very simple to use. Nowadays, every month it is possible to see new features available for cars, which is out of scope of this book.

Other types of Unmanned Ground Vehicles (UGV) for various uses:

In 2016, Domino's Pizza in Australia introduced world's first autonomous pizza delivery robot. The robot is able to choose the shortest path between starting point and destination and can avoid obstacles with its sensors. It seems that autonomous trucks will also appear on our roads soon. On October 2016, Otto and Budweiser completed the world's first shipment by a self driving truck. The truck drove autonomously for 120 miles in Colorado highways, without the driver at the driver seat.

UGVs can also be used as security and surveillance robots, which are discussed under Domestic / Service Robots section.

Unmanned Sea Vehicles (USV) / Autonomous Underwater Vehicles (AUV)

Unmanned Sea Vehicles operate on or under the surface of water, either by remote control or autonomously, for a variety of military, domestic / commercial applications. Designing robots that move over or under water requires special considerations, that must take into account water movements, properties of water, corrosion, pressure, special structural shapes and materials, stability.

Military applications include

- Surveillance,
- Gathering intelligence,
- Attacking enemy ships, USVs or submarines
- Mine detection

Commercial / domestic applications include

- Underwater inspections such as exploring lost ships or even planes
- Cleaning up pollution, such as a great number of small robots acting as swarm
- Scientific applications such as measuring sea currents and temperatures
- Underwater surveys and mapping
- Underwater construction

CHAPTER 5

DOMESTIC / CONSUMER / SERVICE ROBOTS

Domestic robots, help us do daily tasks. Performing various tasks outdoors or around the house, seems natural to us humans, but these are very challenging for robots to perform. Unlike industrial robots, which operate in regular and predefined environments to do repetitive tasks, domestic robots must perform their tasks in partially or entirely undefined or new, irregular and constantly changing environments. These add greatly to the challenges of robot control software and mechanics. Because of this, the field of service robotics, is a relatively new field, in comparison to industrial robotics. Although developing, domestic robots already started to be used and they are doing increasingly challenging, less repetitive, less predefined tasks. Once further improvements happen, domestic robots will be, the real meaning of robots for us, helping us around the house, outdoors, performing tasks to compliment our work or entirely take over what we were doing before, making our lives easier, and probably more fun. There may be robots of completely different sizes and shapes that handle completely different tasks, but in the future if we want a robot that will be able to do everything a human can do, it will inevitably need to be in human form. Below are the main types:

Agricultural Robots

For agricultural products, labor makes up the majority of the costs. Agricultural robots can help reduce labor costs significantly and therefore make a big difference in food prices, which will impact our economy greatly. Because of this, a lot of research is going on in this field, in order to make more capable agricultural robots, such as Sweeper project in Europe, http://www.sweeper-robot.eu/ an EU funded project for making greenhouse harvesting robots commercially available in the market, and many small or large scale firms are also engaged in R&D on their own. The use of agricultural robots will increase greatly in the near future, to the point that it will be normal to see at least one type of agricultural robot on nearly every farm or garden.

Agricultural robots can be used for

- Planting seeds
- Spraying chemicals
- Thinning certain plants
- Controlling and getting rid of weeds
- Picking fruits
- Pruning

As can be seen above, these tasks vary a great deal in terms of the kinds of robots that can perform them, and therefore when we say agricultural robots, it will be many different kinds of robots in various shapes and sizes, ranging anywhere from unmanned tractors to fruit picking robots with highly developed arms, to small weed picking robots in your small garden.

A robot with trimmer attachments, which can also reach hard to reach spots. The robot is remote controlled and can also work on steep slopes with the help of its tracks. Photo Credit: Evatech Inc. - www.evatech.net

Tertill is a solar powered robot, made by Franklin Robotics. It is designed to be used in home gardens, which does weeding by itself. Using its AI, it determines the best time to check out the weeds. It has on board sensors to determine where to go and what plants to spare and what not. A 3D printable design of the robot is also available. Photo Credit: Franklin Robotics LLC www.franklinrobotics.com

Farmbot is an open source farming robot, for small scale food production, made by Farmbot Inc.. It plants at precisely desired locations in any pattern and density, according to user input. It can also detect weeds autonomously, position itself, and with a stroke bury them into the ground. The robot first chooses the suitable end effector, for watering, weeding, soil sensing or planting seeds, and then, by moving on its frames horizontally, and its arm vertically, it can reach to any location within the frames and perform the desired task at the specific location. For watering, it positions itself directly above the plant to be watered, selects water amount according to plant type and age, soil conditions, local weather, thus reducing water consumption. With this robot, different types of plants can be grown in the same area at the same time, and still each plant is individually taken care of. The robot can be controlled by smart phone, through a web based user friendly interface, where user can choose what plant to grow where and the robot does all the rest. The robot is controlled by a Raspberry Pi computer and Arduino controller, and precise action can be obtained through stepper motors and rotary encoders. The hardware and software platform is expendable and all coding is open source. Photo Credit: FarmBot Inc. - www.farmbot.io

Carrier Robots

These robots can help us by carrying our things, by following us autonomously indoors or outdoors, move autonomously in a mapped area.

Photo Credits: Piaggio Fast Forward - www.piaggiofastforward.com

Cleaning Robots

Vacuum Cleaning Robots

It is a robot that cleans the house like a vacuum cleaner. These robots can move around indoor spaces by their wheels and create a map of their environment by digitizing, when they move. While moving they also collect the with their brushes. These are probably the most commonly used type of domestic / consumer robots today.

The most common criteria to look at when buying vacuum cleaning robots include:

- App control through smartphone or tablet
- Scheduled cleaning
- Cleaning area per charge / battery life
- Ability to return to charging station, auto charge and resume
- Side brush
- Remote control ability
- Virtual walls, in order to draw imaginary boundaries for the robot, by the user

New models in the near future is expected to have vision and mapping abilities, recognize objects and even manipulate.

A vacuum cleaner robot example. This robot is called Roomba, made by iRobot Corporation. Photo By: A. Tuter.

Other types of Cleaning Robots

By far, the most commonly used cleaning robots today are vacuum cleaning robots, however there are also other types of cleaning robots for tasks such as:

- Gutter cleaning
- Hull cleaning
- Pipeline cleaning
- Pool cleaning
- Window cleaning

Cooking Robots

Cooking robots which must work with irregular objects that are placed arbitrarily on a table need to have superior visual recognition abilities, dextrous hands with sensors that can measure pressure and position, and a high DOF.

A cooking robot example. Photo Credit: Moley Robotics www.moley.com

Delivery Robots

Delivery robots can make our lives easy by autonomously transporting small items to their required destinations of delivery. Mostly it means the ones operate on land however as discussed before, delivery by autonomous cars or air drones will also be possible. There will be vague or no distinction between a land delivery robot and autonomous car which will be used for deliveries.

Disaster Recovery, Search & Rescue Robots

Disaster recovery robots are used in places where it is dangerous or too narrow for humans to work. These robots can be in many different shape or form, depending on the application. For example, a robot that needs to investigate inside the Fukushima nuclear plant can be a four legged or tracked robot, while a snake shaped small robot can be used inside the rubble, after an earthquake, to detect any signs of life. In 2015, DARPA organized a competition, where several of the world's most advanced robotics teams competed for disaster recovery tasks, such as operating tools, driving vehicles by getting in and out of it, climbing ladder, reconnect a hose, clear obstacles from a doorway and several more. Even though most robots failed at the required tasks, and some call this competition a failure, this was a good start. It must be remembered that 10 years ago, in the same type of challenge for autonomous cars, most cars were not able to complete even half of the track, but we started to see them on our streets.

Lawn Mower Robots

Robots can help with the tedious task of mowing the lawns, by doing it autonomously. They create a map of their surroundings and operate using that map. Below is an example.

Photo Credits: TURFLYNX
www.turflynx.com

Retail Robots

Retail robots help employees and customers in a retail store or a supermarket, with various tasks. They can greet shoppers, scan shelves for products, give customers information about products or directions, make a map of the store and help employees keep track of where items are located. They have cameras, ultrasonic sensors to move around, some can understand voice commands, and for interaction they have on screen displays and even voice. Different concepts such as fully automated grocery stores have also been introduced, such as Amazon's Go store, where a customer walks in, buys and pays with no human intervention and everything is automated through an app.

Telepresence Robots

Telepresence robots can move and serve as an avatar for a person who is not present, by displaying their face and voice on their screen, to facilitate live communication between two people.

Mobile Manipulator Robots

These robots can move around and with their manipulators, act as collaborative robots for a variety of tasks. This is a new type of domestic service robot, they are just starting to move out of factories, and they are the first that can compliment humans around the house for a variety of tasks. These can be considered as early versions of future domestic / consumer robots that will help us in many more tasks around the house.

A mobile manipulator robot example.
Photo Credit: PAL Robotics
www.pal-robotics.com

Social / Personal Robots

These robots are social or personal companions, and can perform useful tasks in stores, exhibitions or homes. The ones for home use can also be called as home robots. They can move around autonomously, detect faces, pets, objects, stream music and video, listen and understand commands, talk, connect to internet, control lights, act as companion for kids, perform security patrol, entertain, and do many more. Many appear surprisingly similar, but do not resemble human form at all, as human form would create higher expectations in our minds that the robot must behave considerably similar to a human. Social robots usually have touch screen for interaction.

Other examples of Domestic / Service Robots

Badminton Playing Robot:
Students of Electronic Science and Technology of China, have developed a badminton playing robot, which is able to play just like any average person. The robot moves with its wheels, can sense incoming ball with its cameras and motion sensors, position the racket accordingly and send it back.

Bricklayer Robot:
Brick laying through robots, can decrease construction costs. These robots can work faster than humans, but still they cannot work on every location and must be loaded with brick and mortar.

Cattle Herding Robot:
A robot to herd cattle was developed by the Australian Center for Field Robotics, in 2016. The robot navigates itself easily through rough terrain, with its widely spaced legs with wheels attached. Future attachments can include temperature sensors to detect animals. The robot also coordinates its operation with drones flying overhead.

CHAPTER 6

HOBBY ROBOTICS

First thing to understand to build robots is the basic logic of sensor (or remote controller) > brain > actuator/motor chain. This is the first thing that should be built, and on top of this, additional components should be added. Building a robot is not easy. An understanding of microcontrollers, circuit prototyping, real-time coding, mechanical design must be obtained. For starters you can go with commercially available hobby robotics kits, or open source, such as very widely used microcontrollers like Arduino or Raspberry Pi or something similar. Also, in _Drones and Robotic Vehicles_ section of this book, the steps to build a multirotor drone was described, to give an example.

Below is a list of some other hobby robot ideas:

- Autonomous indoor mobile robot: With RFID tags placed inside the environment, and ultrasonic sensors to avoid obstacles, a robot can navigate autonomously to any desired location within that environment, even without a camera. Adding a camera will greatly improve the robot's mapping abilities, but it would require much more complex controls.

- Biped robots: Making a biped robot is a relatively difficult hobby level project. In order to achieve walking, the dynamics and stability of walking motion must be studied. There are also many robotics kits around, to build biped robots.

- Gesture controlled robot: In many instances, it is difficult to control a robot with switches or remote controls. Robots recognizing motions can be useful in many applications. The gestures are recognized by acceleration sensor, which recognizes movements in x y z directions.

- Line following robots: One of the most popular and basic types of hobby robots, a line following robot can follow a line drawn on a surface, or even a magnetic field based on what it senses. The robot does this through a constant feedback loop, where it adjusts itself constantly based on line location, as seen by the sensors. This system can also be used for some warehouse or industrial robots, where they are guided on predetermined paths.

- Obstacle avoiding robot with wheels: Probably one of the most basic hobby robots, these robot sense and avoid any obstacles on its path by first sensing them, and then adjusting movement to avoid them. An infrared or ultrasonic proximity sensor can be used to detect objects. When the sensor input indicates that an object is close, as per the distance criteria set by the user, the signals are sent to microprocessor, which in turn adjusts motor speeds to change the turning speed of individual wheels, in order to achieve the steering of the vehicle. This loop is constantly repeated to achieve the ongoing object sensing and steering motion.

- Obstacle avoiding robot with legs: Same concept as above, except legs are used instead of wheels, so servos are more suitable instead of motors. Again, sensors send a signal once an obstacle is detected, and the servo output is adjusted accordingly so that the robot backs off for a certain distance, adjusts its direction and continues walking. Controlling servos especially for walking legs are significantly more difficult than wheeled robots.

- Quadruped robot: Quadruped robots, which resemble spiders in appearance. Usually a quadruped robot will be built with 4 or even more servos. As in biped robots, the forward and backward positions of servos, along with their angles of openings must be coordinated accurately to allow for a smooth walking. As a very basic example, a quadruped can also be built very simply using as little as two servos. To achieve this, two servos can be attached on each end of a bent plate, and to each servo a stick can be attached from its mid point. To each end of the stick, legs are attached. And every time the servos make a move, the legs move, to create a walking motion. Of course the two servos must work in coordination.

- Remote controlled mobile robot: A remote controlled mobile robot can be used for many different purposes, with different attachments such as arms, cameras and a variety of those. For navigation, instead of sensors, it relies on user input. It can be tracked or wheeled.

- Robotic arm: There are so many different approaches to building a robotic arm. Motors, servos can be used for as much degrees of freedom as required, linear actuators, ropes, cables or even pneumatic or hydraulic pistons are possible. An end effector such as a gripper or robot hand also needs to be attached at the tip, for robot arm to manipulate objects.

- Self balancing robot: The robot constantly measures the tilt from vertical, and gives a torque to the robot wheels to compensate for that tilt. This is done many times per second, and from outside it appears that the robot is just steadily balancing itself, without any tilting. The robot consists of basically four components: a controller such as Arduino or Raspberry Pi, an internal measurement unit (IMU), motors and wheels. Two wheeled personal mobility robots work on this principle. Also see *Robotics Terms and Concepts > Ballbot*

Development Platforms

Arduino

Arduino is an open source, low cost electronics platform, for building electronics or robotics projects. It is one of the most commonly used platforms for do it yourself robot projects today. Any type of robot project can be built with Arduino, from a simple line following robot, to a drone that handles its autonomous flight or a complex scientific instrument. It is easy to use for beginners, and has enough flexibility for advanced users. The software that runs Arduino and the board hardware is also open source, the plans of the circuit board is available to anyone, and it can be modified and expanded by users. Arduino has a huge community of users. Main website: https://www.arduino.cc/

Raspberry Pi

This is an open source single credit card sized computer board developed by Raspberry Pi foundation of UK, to aid teaching of programming and computing skills to people of all ages, and has become a very popular platform for electronics and robotics projects. It is a relatively low cost plug and play system, which is perfect for experimenting and runs on Linux. A great deal of free software, applications and downloads are available and it also has a huge community out there. It is possible using Raspberry Pi, just as a computer. Main website: https://www.raspberrypi.org/

Safety points to consider when building robots

This of course is not a complete list, but we decided to open a section about safety here, and list as much items as we can, to raise awareness. Hobby robot building is fun but safety must always be in consideration. About safety, you must consult with any applicable manufacturer, electronics safety guides, someone who has already built the same type of robot, more experienced robot builders, safety manuals, data sheets, consultants, sometimes even authorities such as when operating in public areas, and any other applicable source, before and during building and during operating of any robot as applicable to your situation. Do not take any chances. If you are unsure about something, ask someone competent, and do not build or use it until you know what you are doing.

Here are some highlights:

- Wear personal protection equipment as necessary, which includes goggles (safety glasses), gloves, masks and anything as applicable, not only during building but also during operation.

- Before working on a robot, always remove batteries and / or disconnect any electrical source.

- When working with superglues, make sure you have a solution to debond the glue also, in case it gets on you or any unwanted location.

- Make sure to read all safety guidelines about any battery you are using for your project and frequently inspect them to make sure that they seem in good condition and not damaged.

- Obtain MSDS (material safety data sheets), for any chemicals you are working with.

- Soldering involves hot temperatures. Take precautions to prevent falling of soldering equipment and wear protective gear. Do not touch the soldering iron as it is very hot. Do not leave any hot tools around, after finishing work. Do soldering on fire resistant surfaces and away from carpets and in well ventilated areas. Hands must be washed after soldering.

- High speed rotating motors to be safely attached and operate as they are supposed to.

- Make sure that user manuals for any equipment you use, is in an easily accessible location.

- When working with pneumatic equipment, make sure that the air is vented on all components, before working on the equipment.

- Transport your robot from one place to another in a safe manner.

- Lower all components to the ground such as robotic arms, before working on it.

- Inspect all cords frequently, to ensure that they are in good condition.

- Make sure that when the robot starts, no objects are around that may break, burn, spill etc...

- Inspect all power tools frequently, to ensure that they are in good condition.

- Do not wear long sleeve shirts, especially when working with drills and rotating machinery, as your arm might get caught and serious injury or death may occur.

- Building and working with robots involve electricity. You must follow all applicable safety guidelines regarding electricity, when working with robots.

- Do not plug extension cords, multi device receptacles or power strips into each other.

- Store chemicals in proper containers and with legible labels.

- Make sure that your robot, when starts working, will not pose any safety hazard. Check all screws, loose items, electrical connections, before running your robot for the first time.

- When cutting, direct the knife away from your body

- Do not perform any work while the robot or your hands are wet. Do not turn on the robot or connect it to electrical or battery power source, if any part of the robot is wet, or even if you suspect that it might be wet. Also remember that human body is more conductive when wet.

CHAPTER 7

INDUSTRIAL ROBOTS

The Robotic Industries Association (RIA) defines an industrial robot as a "*re-programmable multi-functional manipulator designed to move materials, parts, tools, or specialized devices through variable programmed motions for the performance of a variety of tasks*"

Industrial robots play a huge role in today's world of manufacturing, for the speedy and accurate production of many items that we use today, by taking over dull, heavy and repetitive tasks from humans. Industrial robots make our lives easier and our production ability greater, but they do not cause more humans to stay jobless. The entry of machines into one field only mean either greater production rate through the use of those machines, or, humans working on more information based jobs to manage the system or new machines, rather than dull, effort based tasks. Industrial robots can also be classified into four groups, such as cartesian, articulated, scara and parallel robots. These terms were explained under *Robotics Terms and Concepts* chapter.

According to IFR, International Federation of Robotics, industrial robot sales are always increasing. Since 2000, with the exception of few years, industrial robot sales have always increased. A total of 99,000 units sold worldwide in 2000, which increased to 248,000 units sold worldwide in 2015.

Today, advanced nations, where labor is expensive, are trying to regain advantage in manufacturing, with the use of better industrial robots, that can perform more tasks, which will decrease outsourcing, as manufacturing with industrial robots are cheaper. If a factory is completely automated for example, even lighting of the space is not needed, which are called dark factories. That alone can improve savings. Industrial robots do not need salary, benefits, or cost anything to work, other than their regular maintenance and the electricity they use. According to an Oxford University study in 2016, 47% of all jobs will disappear in the next 25 years and no government seems to be prepared for this. Of course a good part of these new jobs will be performed not only by what we call industrial robots, but also by domestic service robots as well, which are increasingly taking over more and more complicated domestic tasks.

Industrial robots can be used for tasks including but not limited to

- Pick and place
- Painting
- Welding
- Circuit board printing
- Electronics assembly
- Palletizing
- Testing

- Product inspection
- Labeling

This list is always expanding. For example, a recently developed sewing robot from Sewbo Inc., can handle hardened fabric, by handling it like sheet metal, by moving it under sewing machine at the precise geometry. After that cloth is softened again. Therefore, the task of sewing can now be added to the list above.

For an industrial robot at a production line, or in a workcell for any of the tasks as mentioned above, the most important characteristics include

- The payload it can carry, in other words, handling capacity
- Degrees of freedom or number of axes. To reach points in 3 dimension space, at least three is required but another three axes must usually be added at the end of the robot arm in order to fully control the operation of robot
- Working envelope, which means how far it reaches in space, working range (deg), reach (m)
- Speed (degrees / second) for rotation, horizontal arm, vertical arm and turn, and for linear axes
- Acceleration for different axes
- Pick and place accuracy. Accuracy and repeatability (mm)
- How fast it can do one cycle
- Also, the lower the dimensions of robot, the easier it fits into narrow production spaces
- Load offset (mm)
- Kinematics. For example the joints and bones of your body skeleton define what moves you can and cannot do and it is the kinematics of your body.
- Safety of the robot. Note that even if a robot is certified as safe, it doesn't mean the application will also be safe. This must be separately analyzed for safety. Most industrial robots necessitate restrictions on human entry to the immediate area while the robot is in operation.

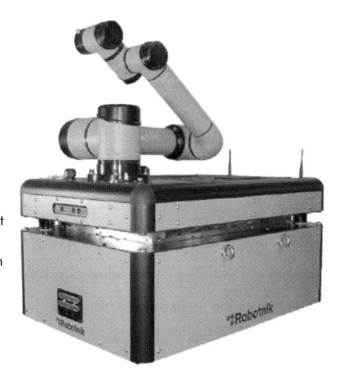

By 2040, more than half of all jobs could be performed by robots. The future generation of industrial robots, will be able to work alongside humans, and be human friendly, in comparison to the cold looking machines that do repetitive tasks

An industrial robot example. This robot also falls under articulated industrial robot subcategory, as it has rotary joints. Photo Credit: *Robotnik Automation S.L.L.* www.robotnik.eu

in an automobile factory. Working side by side will mean that except when human intelligence is needed, all other dull and repetitive tasks a human would do, will be taken over by robots. The new generation of robots, will be able to learn by demonstration, and we already have such robots. The improved abilities and reduced size of newer generation robots will also mean that smaller manufacturers will also be able to use them, which will have a big effect on economy and our way of manufacturing. Increased perception abilities will also allow to perform more repair tasks. A universal basic income for everyone may even be possible soon due to automation. According to a report from United Nations, up to two thirds of all jobs can potentially be taken over by a robot in the foreseeable future.

Also see related terms to industrial robots, under *Robotics Terms and Concepts* chapter of this book, such as, *Accuracy, Cartesian Robot, Articulated Robot, Robot/Work/Automation Cell, Reach, Co-Bot, Degree of Freedom, End Effector, Gripper, Payload, Repeatability, Robotic Arm, Robot Manipulation, Scara Robot.*

An industrial robot arm example.
Photo Credit: *Neobotix*
www.neobotix.de

CHAPTER 8

MEDICAL ROBOTS

Medical Robots can be classified into several categories, such as surgery robots, exoskeletons, robotic prosthetics. Another developing field is the microrobotics, which aims to place micrometer or even nanometer sized robots inside human body, to perform specific tasks.

Surgery Robots

Surgery robots perform surgeries usually with human control, but they can be much more precise than a human hand. Usually extensive training is required for the medical staff and doctors to use these, but once that phase is complete, the benefits can outweigh the cost and time spent. Newer surgery robots are also starting to perform operations on even soft tissues and autonomously, if it is a straightforward task, but of course at all times a surgeon is able to take over control instantly. The robots can also see better than humans by their advanced cameras or even infrared vision. They provide touch feedback to surgeons through *Haptics* technology, which was discussed under *Robotics Terms and Concepts* Chapter.

Exoskeletons

These are wearable electromechanical devices that help people with disabilities to move their limbs, especially people with walking disabilities, eliminating or reducing the need of a wheelchair and providing moving ability that is close to a normal walking motion, therefore increase the quality of life. These can also be used for physical therapy, during the recovery stage of patients who were injured from arms, legs, neck or waist, to regain the ability to move. People who have difficulty in using arms or hands also used specifically designed exoskeleton type robots, for exercise and to improve their abilities. Other types of exoskeletons are more of a domestic service robot, which are worn by healthy persons, or even soldiers, to increase strength.

Robotic Prosthetics

According to World Health Organization, there are 40 million amputees in developing world, but only 5% of those have access to a prosthetic device. This represents a big need and robot prosthetics are now being developed by many companies throughout the world.

Robotic limbs and prosthetics have come a long way. Today we start to see even brain controlled robotic limbs, and not only control but even to restore the sense of touch, as demonstrated by a DARPA research project called Revolutionizing Prosthetics Program, in 2016.

With brain controlled bionic feet, wearer is able to move ankles based on only thought, and the foot has algorithms and sensors that can learn in real time, and automatically adjust the angle of the foot and leg, during different stages as the wearer walks. Not only walking but activities like climbing stairs or getting up from chair are all regulated with sensor activity, which distributes weight for ease of movement.

Robots for therapy

Robots can also be used for therapy of illnesses such as autism treatment. These robots may or may not be specifically designed and manufactured for this task but for example a humanoid robot can be programmed for autism therapy. These can reduce the workload of therapists, by taking over the repetitive tasks, and in some instances recording the performance.

Other medical robot types:

In medicine there are countless of procedures, treatments and operations due to complexity of human body. There are robots that are used and will be used for a good portion of these. For example, robots can now draw blood from your arm, or small micro robots crawl in your veins or intestines for treatment of a variety of things. Personal mobility robots help people to move around, by either balancing on two wheels or serving more like a wheelchair. Nanorobots (nanobots) or micro robots will also find wider applications in the near future, for non invasive treatment of many problems or even enhancing performance of human body.

CHAPTER 9

MILITARY, SECURITY, LAW ENFORCEMENT ROBOTS

Security Patrol Robots

These robots must not be confused with tactical surveillance robots, which we talk about in next subsection. Being a security patrol officer is a tough job. A considerable portion of security patrol tasks are dull, dangerous, physically and mentally demanding and repetitive, which easily puts strain even on people with the highest endurance and attention span. Robots can help with these tasks and help achieve costs savings, as well as improved safety both for the security staff and the patrolled area and act as a force multiplier to security staff. These robots are currently not intended to replace humans but rather to serve as a complimentary force to help security personnel. Security robots are used to patrol inside or around the buildings either by remote control or autonomously, at the same time avoiding obstacles and can automatically recharge, and operate 24 hours a day, detect many things a human cannot and have many other useful features, which can really bring additional value. Like some other subcategories of robots discussed in this book, security patrol robotics is also a relatively new category, but with the advancement of technologies especially in the field of computing, sensors and autonomous navigation, recently there emerged firms and products in this subcategory of robots, which all make robots with different characteristics. As for in each robot category, readers are encouraged to check individual manufacturer websites to see features for each robot. Below you can see examples.

A security patrol robot example.
Photo Credit: Knightscope Inc.
www.knightscope.com

A security patrol robot example. Photo Credit: Sharp Electronics - www.sharpINTELLOS.com

Tactical / Surveillance Robots

Tactical robots are designed for surveillance, negotiation, investigation and at the same time have the manipulation ability with robotic arms and their end effectors. These can be autonomous, semi autonomous or remote controlled. Situations include disposing of explosives, surveillance of an area remotely, hostage negotiation cases, mine detection, and any case where an object must be manipulated or a location must be seen remotely.

This robot is designed to handle dangerous tasks and can operate up to 2 hours, and can be remote controlled in a range of 250 meters. With its double tracks, it can climb slopes of 45 degrees, jump over obstacles up to 20 cm height, and pass over pits of 35 cm diameter. FPV photo and video is possible with the front cameras. Robot made by: ELEKTROLAND www.elektrolanddefence.com Photo By: A. Tuter

Drones are also used for aerial surveillance. Multirotors are usually more suitable for aerial surveillance in domestic tasks. For military applications, plane type drones can also be used.

Weaponized Military Robots and Drones

Today most of the military robots are drones. By saying drones here, as in the definition of drones, we not only mean aerial but also land and sea drones too, so for example a robotic tank or an autonomous military submarine can fall in this category. One thing is clear that robots are increasingly being part of militaries all over the world. As everything else, wars are also being automated and more information technology based. From drones that can throw missiles, to robotic air defense systems that can automatically shoot attackers, to unmanned tanks and boats, miniature spy drones that is smaller than a human hand or the ones that are made to resemble flying insects, tracked mobile robots armed with machine guns, we are increasingly seeing robotic weapons. In the near future, a conflict between two forces may only mean the fight of robots and once resolved, the human fighting force may mean little or even irrelevant.

Robotic systems have many advantages over military equipment used by humans on board directly. Apart from saving lives, computers react faster, and given they have sufficiently strong algorithms, can make better decisions in short time. They also do not have the requirements to protect the human soldier on board. For example, a fighter jet plane with a pilot on board, can make a turn which would create an acceleration up to a certain allowed amount, in order not to harm the pilot, but a robotic fighter jet will not have this limitation and can make sharper and faster turns with higher acceleration value, which is a clear advantage. A fighter jet with a pilot also has a lot of different features to protect the pilot at all times or even eject him/her in case of emergency. For robotic fighter jets, this will not be necessary, which can cut costs and make higher performance jets possible.

Increase of usage of robots in military will also bring the necessity to support this with a whole new information technology and cyber force division for all militaries, so in addition to the traditional land, air and sea branches of militaries, we are already starting to see a cyber / information technology branch. Given the ever accelerating development of information technologies, it is possible that in a few decades we will have military forces that are unimaginably different than today, from swarm of very small drones to giant fighter robots, to androids that can carry guns which are multiple times stronger than humans with their artificial muscles or actuators. Discussing military equipment however, is not included in the scope of this book and we will leave it here.

CHAPTER 10

SENSORS

Sensors are devices that detect or measure physical properties, by converting form of energy into electrical signals, so that these inputs can be used in electronic devices and systems to produce output, or certain behavior, and they play a major role in all of our electronic devices today. Therefore, in a book about robots, it is necessary to talk about them.

The important criteria for sensors include:

- accuracy
- calibration
- limits of environmental conditions to operate
- resolution (smallest increment of property that can be detected, in other words, sensitivity)
- response time
- current consumption
- sensing range
- data noise

Sensors provide useful information about the surroundings, or the internal workings of a system or machine. They can be classified mainly into two groups, as passive and active sensors. Passive sensors do not release energy into the environment, such as a camera. Active sensors release energy into the environment in order to receive input. For example, ultrasonic sensors must release sound waves into the environment, before it can measure the reflections. Output from sensors can be analog, binary or digital. Sensors must be calibrated before any measurement. A good quality sensor should not measure unnecessary input but must be sensitive to the property it is measuring. It also should not have an effect on the property it is measuring, in other words, it should not cause any change for anything it measures.

A sensor has basically two components. The interacting element interacts with the environment by sensing it, and provides information via causing a change, for the other element, which is the transducer. A transducer is an instrument that converts energy from one form into another, such as variations in a physical quantity into electrical signals, such as a microphone converting sound pressure into electrical energy, or a motion sensor converting infra-red energy into electrical signals.

For a robot, sensors provide useful data about the robot's environment, but they can also measure the internal state of any component or components of the robot, while it works. Things that can be measured with sensors include but not limited to the following:

- Force
- Temperature
- Light
- Distance
- Orientation
- Rotation
- Acceleration
- Electrical Charge
- Voltage
- Humidity
- Pressure
- Human Input
- State of a machine component
- Sound
- Properties of a material
- Magnetic waves
- Vibration or shock
- Location
- Speed
- Linear or Rotational position

Basically anything that affects a machine for its operation, as well as the control the inner workings of that machine, are measured and quantified by sensors.

Sensor fusion, is combining of sensory data in order to have more accurate and better information than the total of separate inputs, therefore creating synergy, and also to measure something that would not be possible by using only one sensor alone or using sensors separately. In other words, by combining data from sensors, the application or system performance is improved by creation of a sensing platform. Combining data from different sensors, from multiple data sources, by leveraging capabilities of each sensor, and fusing them together via sensing algorithms, corrects for the errors of individual measurements, therefore increases data confidence, reduces uncertainty, improves resolution, enables data to continue to flow even in case of an individual sensor failure, increases spatial coverage that is being measured, reduces system complexity by combining data from sensors into one smooth source of information instead of multiple sources of less reliable information. The sensors can be of the same or different types. A good example is combining of image data from two cameras, to get a 3D view, just like we have two eyes, to get a 3D view of our surroundings. For example, an accelerometer, gyroscope, and a magnetometer input can be combined, to control orientation and acceleration of a drone, or in cell phone, through MEMS sensors and fusion of data, screen rotation, step counting, navigation, gesture recognition, gaming can be performed. Various statistical methods are used, to process these data, such as Kalman filtering, Bayesian networks.

A sensor network is a group of tiny sensing devices connected to each other wirelessly in order to monitor conditions in an environment. As the data about the environment is collected, it can be used in a number of ways, including robotics applications, although the primary purpose to date has been for monitoring purposes.

Below we list and introduce major types of sensors, but remember that there are hundreds or even thousands of sensors, if you also consider the subcategories of these sensors and other types of sensors and their subcategories, that are not listed here due to being rather less relevant. For all of the main sensor categories we list below, there are many subcategories, made differently, in different sizes and weights, working differently and measuring things for different purposes, in different situations with very wide price range and sensitivity. Volumes of books can be written about sensors, but we tried to limit it to the most relevant ones for the purpose of this book.

Acceleration sensor (Accelerometer)

It is an electromechanical device that measures acceleration due to increasing or decreasing speed, fall, vibration, shock, or high g turns. It can be 1,2 or 3 axes, although for most robots and for all drones, 3 axes accelerometers are used. Measuring acceleration is an important aspect of robot control. For example in drones, by measuring acceleration constantly due to the effects written above, the drone can be stabilized with constant feedback loop to the motors. Accelerometers work based on a basic physical law, Newton's second law of motion, which states: Force = mass x acceleration. Therefore, when there is a known mass, and if we can somehow measure the force that acts upon that mass, we can calculate the acceleration. There are different types of accelerometers based on how this force is measured, but they all work using this basic principle.

Types of accelerometers include:

- Mechanical accelerometer, which is the most basic type, measures acceleration simply by attaching a mass to a spring, and measuring the force by the distance that spring travels, when the mass is attached. The constant of the spring (k) is known, and based on the spring equation of F=k.x, force can be calculated once the distance (x) is measured as a result of an outside force. Once this force is calculated, from F=ma the acceleration can simply be calculated by dividing F by m. These types of accelerometers obviously are too large and rather used for calculating the magnitude of an earthquake for example, in seismographs.
- Capacitive accelerometers, works based on the principle that the capacitance change when the distance of plates of a capacitor change. Therefore, when the mass moves, the capacitor has different amount of electricity, which is then translated into measuring of acceleration.
- Piezoelectric accelerometers use piezoelectric crystal material. When the mass moves, it presses or pulls away the material, which produces different amount of electricity. This is then translated into acceleration information.
- Semiconductor chips, are the type of accelerometers that are used in electronic devices such as smart phones, robots, automotive industry, home appliances, medical devices and more. They essentially work with the same principle of capacitive accelerometers, but they are in much smaller scale with very tiny moving parts and capacitors. They are called MEMS, which stand for micro electromechanical systems.

Accelerometers are considered a sub group of inertial navigation sensors together with gyroscopes. Accelerometers, gyroscopes and often times magnetometers can be combined in one sensor called inertial measurement unit (IMU) and connected to flight controller in drones or navigation controller for robotic vehicles. With IMU, in the absence of GPS signals, the position can still be calculated (of course this method is exposed to accumulative errors, but still provides something, in the absence of GPS) with a concept called dead reckoning. The acceleration calculated from accelerometer can also give the velocity with respect to previous velocity, since the derivative of acceleration with respect to time is velocity. A clock calculates the time, and velocity times time is distance, and this way, dead reckoning can still calculate the distance. Also see *Inertial Navigation System, INS, and Inertial*

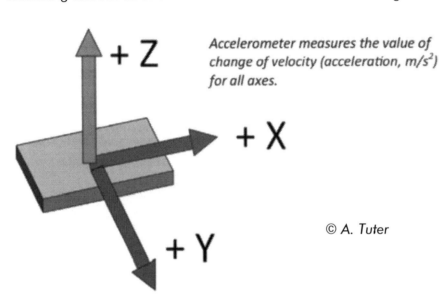

Accelerometer measures the value of change of velocity (acceleration, m/s²) for all axes.

© A. Tuter

Measurement Unit (IMU), and Drones & Robotic Vehicles > Unmanned Aerial Vehicles > Terms in relation to Drones > Dead Reckoning section.

Air Pressure Sensor

Air pressure sensors (barometer) determine the height of a drone, based on the principal that the air pressure decreases as height increases.

Altimeter

This is for measuring the altitude of the drone, with respect to sea level. Altimeter can also be used to stabilize the height of the drone, when the operator sets it to hover mode.

Camera

Cameras play an essential role in robot or drone operations. There are countless types of cameras used in robots for countless of different purposes.

For example the most frequent use of aerial drones is to record videos or take photographs, therefore cameras are very important components of drones. Cameras may come with the drone, as built in cameras, or they may be external that can be attached later. External cameras give more flexibility and control about the camera selection and they have their own batteries. The compatibility of camera and drone and the user's photo and video needs determine which type of camera or drone to choose. A camera with a remote shutter is also desirable. Some drone and camera systems come with phone apps that can be downloaded to a smart phone and managed directly from phone.

For other types of robots, cameras also come into the picture very often. For example, an industrial robot can have a camera attached to the end of the arm, in order to perform operations based on visual input, or a tactical robot, can use a camera for remote surveillance. In advanced fields such as defense sector, events broadcasting, or robotics, the ability of cameras to track very fast moving objects, or even tracking multiple objects with a single camera becomes very critical.

Cliff Sensor

Cliff sensors prevent robots from falling when they reach to an edge, such as the edge of a table, or a stairwell. They are commonly seen in domestic / consumer robots such as vacuum cleaners. They can be optical, mechanical or ultrasonic. An optical cliff sensor reflects a beam of infrared light on to the ground surface. When the robot reaches an edge, the reflection received changes its form, which signals the robot to stop and move in the other direction. A mechanical cliff sensor works the same way, except that there is a thin mechanical piece that contacts the ground continuously at the robot's sides. When an edge is reached, the position of the contacting piece changes, which again prompts robot to stop or change direction. Ultrasonic cliff sensors are the same as optical ones except that they reflect a sound wave, instead of a beam of light.

Electro Optical Sensor

These are sensors that convert the quantity of light into electrical signals. These are always part of larger systems and used in a large variety of applications, wherever the light must be converted to signals, from position sensors, to smart phone screens.

Force Sensor

Force sensors are used to measure change in force or rate of change of force. This way, they can also be used for touch detection. There are many different types of force sensors. Strain gauge force sensors, capacitive force sensors, resistive force sensors are the most common. For electronics applications, capacitive or resistive force sensors can be used. For a capacitive force sensor, the capacitance changes when a force is applied as the distance between capacitor plates changes. This way the force is measured. For a resistive force sensor, the resistance of the pressed material, such as a

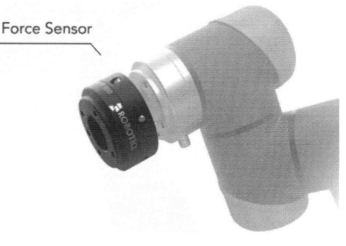
Force Sensor

Photo Credit: Robotiq - www.robotiq.com

piezoresistive material changes, and this way the change of force can be quantified.

GPS Sensors

GPS sensors use satellites to provide an accurate position on earth. GPS satellites constantly broadcast their position and time by radio signals. A GPS receiver uses these signals and determines its position anywhere on earth. The receiver must see at least 4 satellites, to determine accurate positioning. GPS sensors are very widely used today, for any mobile electronic device, such as cars, cell phones, and in many commercial, industrial and military applications.

Gyroscope

Together with accelerometer, gyroscopes (also called "gyros" in short) are considered a sub group of inertial navigation sensors. They can be of mechanical, optical or vibratory type. Mechanical gyroscopes work by measuring the amount of rotation, by mechanically isolated wheels. Optical gyroscopes send light in both directions and when turning occurs, a phase difference between the two beams can be measured. Vibratory gyroscopes measure forces that act on a resonator based on its direction.

Gyroscopes are a part of inertial measurement unit, IMU, which is a component of inertial navigation system (INS). They measure rate of rotation about an axis, in degrees per second, but usually measurement about all 3 axes is needed, especially for drones. Therefore the term "3 axes gyroscope" means, that gyroscope is able to measure rotation in all possible three axes, pitch, yaw and roll, which were discussed under *Terms in relation to Drones* section. They are a fundamental component of motion control, sensing and stabilization for all robots and drones.

Hall Effect Sensor

When a conductive material goes through a magnetic field, it produces voltage called Hall voltage. Hall effect sensors uses this principle to measure magnetic field intensity, or in other words, magnetic flux. A Hall effect sensor is a type of magnetometer. Detection of magnetic field helps detecting position, speed, proximity sensing, current sensing in a wide range of automotive and industrial applications. Even inside a brushless motor, motor controllers need a Hall sensor signal in order to determine the rotor position. Some of the main advantages of Hall effect sensors are, they are not affected by environmental factors, such as air, dust, dirt, radio frequency noise, and have stable sensitivity in varying temperatures.

Heat Sensors

See *Temperature Sensors.*

Inclinometer

See *Tilt Sensor.*

Inertial Navigation Sensors

Inertial navigation sensors are part of inertial navigation systems, as was discussed under *Inertial Navigation System, INS, under Drones & Robotic Vehicles > Unmanned Aerial Vehicles > Terms in relation to Drones* section. For example, accelerometers and gyroscopes are inertial navigation sensors. Linear and angular accelerometers and gyros provide measured accelerations and angular velocities to a computer which calculates direction, speed along that direction, and integrates that to obtain an estimate of its position, as in *Dead Reckoning*. However, without the use of GPS, errors will accumulate and the estimated position will drift gradually further from true position, so GPS updates at certain intervals are used, but this is not a part of INS. For measurements of inertial sensors, *Kalman filtering* is used, which is an estimation algorithm for producing more reliable results from the measurements of the sensor. For robotics projects, auto-balancing, auto-rotate-to-angle, field-oriented drive, collision detection, motion detection, are some of the aspects to consider when choosing INS. Some of these are open source, such as with firmware source code, board schematics/layout & bill of materials available online. Below is an example.

Combined with the ping2020 Class A1S transceiver to provide a complete ADS-B IN/OUT solution

Navigation Sensors, and their connection to an autopilot.

Photo Credit: UAVionix — www.uavionix.com

Infrared (IR) Sensors

Infrared sensors can detect and measure infrared light. An infrared beam of light is emitted first, and when this beam of light is reflected from an object, the IR sensor picks it up. This is useful in many applications, from line following robots, night vision goggles, to remote control of our televisions.

LIDAR

LIDAR stands for Light Detection and Ranging, and measures distances with the help of pulsed lasers. By using many pairs of laser beam reflections which are sent in high frequency, to all directions, the distances to any objects or natural features in surrounding area can be measured easily, in high resolution and a three dimensional map can be gathered quickly as well as positioning the robot. It uses the same principle as a radar, but the wavelength here is 100,000 smaller than radio wavelengths and therefore producing maps of much higher resolution is possible, which is an important quality for use in robots, because robots need to identify small objects as well, while navigating around them or manipulating them. This is usually achieved by rotating the source of laser 360 degrees and in various angles or heights and calculating the time that a certain beam is reflected from a certain point in 3D space, in order to calculate the distance of that point to the source. Repeating this in high frequency for thousands or even millions of points produces a high resolution map of the surrounding space. For this reason, LIDAR is also called 3D scanning or laser scanning. Since laser can hurt human eye, safety of eyes is also a concern when building a LIDAR system. In more advanced LIDARs, in addition to distance, additional properties of the surrounding objects can also be measured, such as the velocity or material composition, by making use of the Doppler shift of the reflected light. Main components include laser emittor, laser detector, rotatable housing, scanner and optics, inertial measurement unit, GPS, mounting base.

Critical factors for LIDAR include:

- 3D point density that it can produce, such as 100 or 200 points per degree around each axis,
- angular resolution in degrees, such as 0.01 degrees angular resolution,
- distance to which it can reflect its laser beams such as 200 meters,
- sampling rate in Hz, such as 25000 Hz

As of the date of this writing, commercially available LIDAR systems are still somewhat expensive, ranging from a minimum of around $1000, and can go up to more than fifty thousand dollars. However constant improvements are being made to make the sizes smaller and reducing costs. This will eliminate a huge obstacle for robotics and autonomous vehicles since they heavily depend on LIDAR in order to navigate and map surroundings. LIDAR is widely used in various robotics and other applications today including on autonomous or remote controlled land, sea or air drones, autonomous cars (also often referred to as self driving cars), many types of robots that operate in commercial or industrial environments, in many fields of science and industry for the purposes of mapping or navigation. Examples other than robotics use include mapping, construction industry, where an indoor 3D map is quickly created and this is used for coordinating dimensions of materials and installations with each other, agriculture, infrastructure, forestry, geomatics, archeology, gaming, geology, oceanography, military, mining and more.

Light Sensor

Light sensors detect light and create voltage difference in proportion to change in light intensity that falls on to the sensor. *Photovoltaic Cells, Photo Resistors, Infrared Sensors* are different types of light sensors and each are explained under this section.

Magnetic Sensor (Magnetometer)

Magnetometers measure the magnetic field strength. They are also used for detecting the compass position of a robot. The flux density of a magnetic field is measured in tesla units, which is newton meter / ampere. Magnetometers can be used in conjunction with accelerometers, and gyroscopes, to measure compass heading and can be a part of inertial measurement unit (IMU), which is a very useful input especially when calculating position with dead reckoning, in case there is absence of gps satellite signals. In drones, apart from positioning, magnetometers can also be used to detect different minerals underground, or metallic underground piping. Also see *Drones > UAV > Terms in Relation to Drones > Dead Reckoning*

MEMS Sensors

MEMS stands for micro electro mechanical systems. Building electro mechanical systems in micro sizes enable placing them in small sensors which can be placed in compact electronic devices or systems that we use today. Accelerometers, motion sensors, gyroscopes, pressure sensors, tilt sensors, may all use MEMS technology, depending on the type. Use of MEMs technology enables things such as screen rotation, step counting, navigation, gesture recognition, flight control and stabilization, gaming, and many more functionalities that various electronic devices perform.

Motion Sensor

A motion sensor, also called motion detector, is a sensor that detects moving objects, as most frequently used in building security systems, to detect people. Mainly they can be infrared, microwave, or ultrasonic type, which emit and reflect these waves. They are used for robots which need to respond to moving objects or humans.

Multimeter

A device that can measure voltage, electric current and resistance.

Navigation Sensor

See *Inertial Navigation Sensor.*

Photodiode Sensor

A photodiode converts light into electrical current, which is used to measure the light intensity. It is a subcategory of photodetectors. They are used in high speed photo detection, copiers, gaming platforms and many types of robots where light detection is required.

Photoresistor

A type of light sensor. A photoresistor changes its resistance based on light intensity, and therefore act as sensor.

Photo Voltaic Cells

A type of light sensor. Photovoltaic cells convert light and solar energy into electrical energy, and in this sense, they are more of an energy source, than a sensor.

Position Sensor

Position sensors measure the distance travelled by the measured body, such as the robot itself, in reference to a starting point, therefore it is an internal sensor. Linear or angular distances can both be measured. They have a lot of applications such as packing robots, medical equipment, drive by wire cars, military equipment, in other words, in any machine where the position of components need to be measured.

One of the most commonly used type of position sensors is a potentiometer, because of its low cost and ease of use. A potentiometer has a contact to a mechanical shaft that is either angular or linear, and when the contact moves, it comes to a different position on the shaft, which is a resistor, so the resistance value changes with the changing position. The linear progress or angular progress can be measured as a distance or angle by the current that goes through the varying resistance. A major limitation of this is that the position measured is limited to the size of the linear or angular shaft. Therefore, this is suitable for use in applications where only small changes in position are needed to be measured, such as the needing to know the angular position of a robot arm, or the small linear distance covered by a machine part while in operation.

A capacitive position sensor measures the position either by changing the dielectric constant between the plates or by changing the overlapping area between plates, which is a result of changing position that is being measured.

There are other types of position sensors such as, magnetostrictive, inductive, optical or hall effect position sensors, with different uses.

Potentiometer

A potentiometer is a variable resistor based on a moving terminal, through external mechanical action. When the moving terminal changes its position on the resistance, the resistance value changes. It can be used as a position sensor or in an application where variable resistance based on an external action is required for any reason.

Proximity Sensor

Proximity sensors can be mainly two types, as infrared or ultrasonic. They transmit infrared light or ultrasonic sound waves, which is then reflected from objects and based on the time elapsed, the distance to any obstacle or nearby objects can be calculated.

Sound Sensor

These are basically microphones, which helps a robot to respond to sound inputs.

Spectrometer

A device that is used to capture different colors of light. Distinguishing colors is important in a wide range of applications such as oil and gas, forestry, agriculture, environmental monitoring industrial automation.

Tactile Sensor

See *Touch Sensor*.

Temperature Sensor

Heat or temperature sensors can detect temperature differences for many different reasons. They can even be used for locating people after a disaster for a search and rescue robot for instance. They produce a voltage output linearly proportional to temperature increases.

Tilt Sensor

With reference to gravity, tilt sensors measure tilting position of an object. They can also be called rolling ball sensors, mercury switches, inclinometers or tilt switches. They can be one or two axes for robotics applications or 3 axes for gaming applications or in cell phones. They measure the tilting angle by producing a variable electrical signal due to varying inclination.

Touch Sensor

Touch sensors replaced buttons and switches to eliminate the problem of mechanical wear and tear. They can contain micro mechanical switches, such as a mechanical based sensor, or a resistive material that changes its resistance upon touch which is a resistive based sensor. They are closely related to *force sensors*.

Torque Sensor

A torque sensor measures the amount of torque, such as a static torque, on a static, not moving piece or a dynamic torque on a moving shaft. They can also be called as torque transducer or torquemeter.

Ultrasonic Sensor

Ultrasonic Sensors work by reflecting sound waves from surfaces which enables measuring distances and therefore positioning a robot. On a drone, these must be placed as far away as possible from noise or vibration sources such as propellers. For example they help a robot to avoid obstacles or are used to position a drone usually in indoor operations.

Vibration Sensor

Vibration sensors are used in robots, to monitor the smoothness of operation by detecting vibrations.

Visual 3D Depth Sensor

Two elements must work together in order to detect 3D depth, an infrared (IR) projector and an infrared camera. The infrared projector emits infrared light to the environment and the IR camera catches this light. An IR camera is essentially a regular camera with only difference that it works with IR light. The camera sends the light, which is in the form of reflected dots from pointed light beams, to the processor. The processor than analyses this data, so that the points that are closer to the camera have more dispersed reflected dots, and the points that are further from the camera have more dense reflected light dots. This enables the depth calculation. Of course IR light it is not visible to human eye. Another type of depth sensor is a stereoscopic depth sensor, where two cameras are placed side by side, similar to a human eye configuration, in order to measure depth. This is made possible by comparing the images from the two cameras, which essentially is the same thing our brains do when we look around, and that is the reason humans cannot sense depth if one eye is closed.

An example of a stereoscopic depth sensor. Two cameras, which are placed at different viewing positions, are able to catch 3D depth of the scene, by combining images, which works with similar principle to human eyes. Photo Credit: Nerian Vision Technologies www.nerian.com

Voltmeter

A device that is used for measuring voltage, in other words, potential difference between two points on an electric circuit.

CHAPTER 11

SOFTWARE & PROGRAMS

Robots have computers that control them and like any computer those computers also need code and software to operate. Common tasks for a robot software includes control, feedback loops, navigation, locating objects, data filtering and sharing.

Robot control can basically be divided as follows:

- Perception or sensing with data obtained from sensors,
- Modeling of the environment and the robot itself,
- Planning the mission, and the actions required to do it,
- Executing planned actions by the control of motors and actuators

A robot software architecture may be constructed as follows:

- Managing user interface, sensor data display, robot status, image displays
- Running high level algorithms, such as navigation and mapping, planning tasks
- Code that translates between low and high level algorithms, such as sensor fusion, steering, kinematics, image processing
- Running low level code for sensor and actuator interfaces

Note that each robot will need its unique software architecture.

Computer Languages & Programs for Robots

The programming languages you see below are mostly general purpose languages that have worldwide uses, for many industries as well as robotics. For robot control, these languages also form the basic building blocks to hobby level or industrial/commercial or advanced level robotics. Before we talk about any of the languages below, for someone who is not familiar with computer programs, we must mention the very basics: There is the machine language that only a computer can understand, and then there are low and high level languages that humans write. The lower level the language, the closer it is to the machine language. The higher the level, the closer it is to human language and more user friendly. Both have certain advantages and disadvantages.

Arduino IDE

This is a specific development platform and programming environment for Arduino, which is one of the most popular open source robot controllers today. All programs to control an Arduino based robot can be written in Arduino IDE, which is C/C++ based. The IDE is first installed on PC, Mac or Linux, and then, the program is written in C++, and finally it is downloaded onto Arduino board. With Arduino IDE, simplified commands can be used, or the microcontroller registers can be changed, or just C language can be used. https://www.arduino.cc/en/Main/Software

C

C is a low level, general purpose computer language which is fundamental for dealing with embedded computing and hardware in countless industries and applications today. In many ways, it can be considered as the primary language of computer languages. The fastest code, which will need the least sources from the computer, can be written with C, because it is low level. In other words, C does not require a lot of interpretation to machine language, it deals with the hardware directly. In fact, C is probably the lowest level general purpose language, which is still computer architecture independent. You go any lower than C, and your language must now be dependent on the architecture of hardware. We need to clarify that, in comparison to those hardware dependent languages, like Assembly or machine language, C is of course higher level, but those machine level languages are only used by the people who directly deal with computer hardware and we are talking about general purpose programming languages here. C is lower level, and therefore more fundamental, than any other languages mentioned under this section. Therefore C can be considered essential for a true robot enthusiast, but, being a low level language, it may be difficult to apply to more complicated robots, where C++ comes in as discussed below. Small hobby robots often use C or a custom variation of C. Most microcontrollers are programmed in C for instance. C may not be a friendly language for a beginner as it is detailed and not so user friendly.

C++

C++ can be considered the most essential language to know, for anyone who wants to work on more complex robots. Simple robots can be done by C, but when dealing with the complexities with more advanced robots, the extra features and power of object oriented programming of C++ would be more useful. C would be too bogged down on details for more advanced robots, being a lower level language. For example, consider an industrial robot, which works on a complicated curved shape, that is made of many curves. To manage movements of the joints of the robot accurately in 3d so that it can work on (spray paint for instance) all of these surfaces properly, the program must handle equations of curved planes, matrices, vectors and coordinates in 3d. C++ is much better suited to do these than C, with its power of classes, objects, operators, math libraries. Most robotics simulators are written in C++, then followed by Python, and then few other languages.

Note that it is also common for control logic for microcontrollers to be written in one language, and advance logic for things like image processing or machine learning be written in another language. Whatever language is used, it needs to have good support for the particular hardware on the robot, as well as effective real time characteristics and good support for the protocols and networks that the robot will use to interact with the real world. For example C language can be used for situation with tight real time characteristics and limited memory, such as to control specific peripherals, microcontrollers, where a more user friendly and higher level language can be used to deal with a user interface component, or making your robot to do more complex things, when a low level language like C will remain too bogged down at the hardware level. For a humanoid robot, the motor controls might be written in C, scripting can be done in Python or Javascript, and image processing can be done in C++. C# can be used for speech processing, and the overall behavior system might be programmed in C/C++ or Java. ROS, RobotC, Arduino, Rapsberry Pi are few of the examples where C++ is used, with many more, just in robotics.

Java

Being one of the most popular computer programming languages, Java can be used in many different types of robots, ranging from hobby robots to industrial robots. Java is a high level language and does not deal with memory usage like C or C++, and therefore can be suitable choice for performance and ease of use.

LabVIEW

LabVIEW stands for "Laboratory Virtual Instrument Engineering Workbench" and is made by the company National Instruments, headquartered in USA. It is a graphical programming language, which is one of the most advanced and widely used development environment to build largest and most complex applications in the world, for data acquisition, test, measurement, instrumentation, control, and analysis applications, by scientists and engineers to use for rapid design development and deployment. Here is the main website: http://www.ni.com/labview/

For robotics applications, it can be used for sophisticated projects which include sensor communication, obstacle avoidance, path planning, kinematics, steering, and more. It can be used for implementing mobile robot platform layers and system design. It has no hardware or software prerequisites. It has libraries for many widely used systems and platforms.

MATLAB

MATLAB, is made by The MathWorks Inc., headquartered in USA, is a widely used and advanced high-level language and interactive software environment for numerical computation, visualization, and programming. It can be used for many applications such as machine learning, data analytics, internet of things, motor and power control, wireless communications, robotics, mechatronics, test and measurement, computational finance, image processing, computer vision and more. https://www.mathworks.com

For robotics, it can be used for designing and prototyping robots, test algorithms, connection to robot peripherals and platforms. It also has a good source of sample projects, which can serve as very useful references for users, which show sample codes and logic construction for a lot of (as of writing of this section, we found 52 on their site), different types of robotics projects which included projects such as controlling a robot arm, virtual design tool TurtleBot for ROS, path planning robot in environments of different complexity and many more different types of projects. https://www.mathworks.com/examples/matlab-ro. After a prototyping is done in a relatively easier way than for example C/C++, then this can be reimplemented in C/C++ as necessary, since C/C++ is a lower level language and things can be rewritten to use resources more efficiently.

Python

Python, is one of the most commonly used programming languages for robotics, is a higher level language than C or C++, therefore it is more user friendly but further from the hardware. Python is easy to use, it still interfaces with C / C++, has lot of robot compatible libraries, such as the Open Motion Planning Library (OMPL), which is really time saving. Python is a core language of ROS (Robot Operating System), which means it is fully compatible with ROS and the full power of a distributed robotics system and all its libraries/tools are available via Python. For example, using Python can enable faster development of a motion planning library than C++ development, but then in the end, a final version can still be written in C++, to make it more efficient as C++ can better manage memory, better suited to run things that need to run in higher frequencies, being a lower level language than Python. Faster development with Python also enables faster experimenting with simplified programs, therefore it is more suited to prototyping and it also works just fine for hobby robotics. For more advanced commercial or industrial robots, it may not be the best choice, since it may not provide the speed and reliability required in those situations. https://www.python.org/

Raspberry Pi

Like Arduino, this is one of the most popular open source development platforms and has a huge user base. To program and run Raspberry Pi, all you need to do is to download the operating system into your computer, and then with the help of a special program you install this Linux based operating system into an SD Card through your computer, which is then plugged into Raspberry. At this step the operating system is installed. After this, you can make configurations based on your project. The operating system is called Raspbian, https://www.raspbian.org, which is run by a separate organization than Raspberry but still free to use. It is a form of Linux and has all the tools needed to program with, such as 35,000 packages, pre-compiled software bundled in a suitable format and directly installed on Pi. Raspberry Pi is actually a full computer that runs with a full operating system. The programs are written within that operating system, and run on Pi, just as any other computer, and therefore no need to write anything on your computer but it is still possible to do. With the operating system, there is also access to the online repositories to add support for almost any other programming language that exists. The most commonly used languages include Python, C/C++, Java and almost any language supported by Linux can be used (and that is most of them) and is already included in the OS. Main website: https://www.raspberrypi.org/

Robot Operating System (ROS)

There are many open source robotics software out there today. Open source has the main advantages as, ease of sharing information, prevention of repeating previously worked and resolved items by different users, which all leads to faster, more economical and effective development.

Today probably the most commonly used robotics software is ROS, which stands for Robot Operating System. From the start it was developed by multiple institutions with multiple robots, including many which received PR2 robots from Willow Garage Inc. based in USA. This is an open source software, and has a huge number of users involved. It contains set of libraries, algorithms, developer tools and drivers for developing robotics projects. The first release of ROS was in 2010, and as of beginning of

2017, ROS has reached its 10th official release, which is called "ROS Kinetic Kame". There are translations to 11 languages other than English, which are: German, Spanish, French, Italian, Japanese, Turkish, Korean, Portuguese, Russian, Thai and Chinese. It currently has 2000+ software libraries, which keeps increasing every year. http://www.ros.org is the main website and it is being maintained by Open Source Robotics Foundation.

Many robots use ROS now, including but not limited to hobby robots, drones, educational or advanced humanoid robots, domestic / service robots including cleaning robot vacuums, cooking robots, telepresence robots, robot arms, farming robots, industrial robots, even Robonaut of NASA in space or the four legged military robots in development. A list of robots which use ROS can be found here: http://wiki.ros.org/Robots. In this list, the link to each robot shows varying information such as packages, installation, resources, related software if any, usage and other details about how ROS is used for that robot. It virtually standardizes the robotics software and therefore prevents "reinventing the wheel" by individual robot developers. Bottom line of what we can say about ROS is that, it seems that this open source robotics software, is one of the most beneficial tools to anyone beginning or involved in robotics. It seems good idea that anyone who starts in robotics, should start with ROS.

We were checking the Alexa.com (a website that is owned by Alexa Internet Inc., a subsidiary of Amazon.Com Inc., which also gives you the ranking of websites in the world) ranking of ROS since few years, in order to track the increase in usage. On the left are dates we checked and the numbers on the right indicate the ranking of ROS.org website from top, among all websites in the world:

As can be seen here, in 2011, when we first checked this ranking, ROS.org was at 189,000 th place in the world from the top among all other websites in terms of unique visitors that visit the site, and it almost continuously increased its ranking. As of February 2017 when this section was written, it was

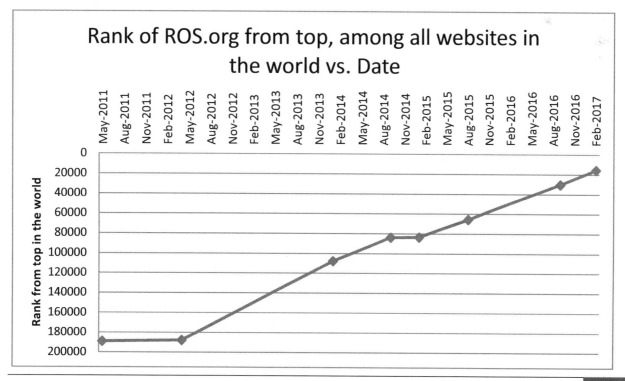

the 15,621st most reached website in the world, with mostly being accessed in Japan. The ranking of 15,621 means ROS.org is a very high traffic website.

The percentage of reach among countries for ROS.org is as follows as of February 2017:

Japan 22.9%, China 19.8%, USA 13.7%, South Korea 9.0%, Germany 5.2%

Author's note: This was checked again on August 2017, just before the time of publishing of book, and overall ranking of ROS was 14,968 from the top, which suggests that the increase in usage is continuing, and it also suggests that the overall user base is now established at a very high number.

Date	World Ranking	Top Country
May-2011	189000	no data
Apr-2012	187900	no data
Jan-2014	107821	no data
Sep-2014	83875	Canada
Jan-2015	83556	Canada
Aug-2015	65754	USA
Sep-2016	30201	China
Feb-2017	15621	Japan

On ROS website, in addition to all packages, there are also extensive tutorials and a discussion board that one can ask questions and share knowledge. The discussion board can be found here: http://answers.ros.org/questions/

ROS also has an industrial section, the version of software modified for industrial applications. It is called ROS Industrial, and can be reached at: http://rosindustrial.org/. Although we see domestic robots with new abilities or advanced research projects that aim to develop capabilities of robotics every year, according to the results of a study that is shown on http://rosindustrial.org/the-challenge/ website, the abilities of industrial robots are not progressing and the abilities are restricted to welding, material handling, dispensing, coating (although we know that they do additional tasks such as packaging, inspection, labeling etc...). ROS Industrial aims to solve this challenge by providing a common skeleton to all developers, with its extensive and stronger software architecture, in comparison to other individual robotics programs.

ROS is also a communications protocol that lets different devices that run with different languages, APIs, timings, etc... to talk to each other. ROS makes possible to program a robot's navigation and limbs, regardless of what hardware is used. Therefore it is a highly promising tool to standardize operations of robots all over the world. It lets people use components such as cameras, motor controllers, encoders, PCs, etc.. and have them work together without entering too much into the various interfaces to create a custom framework from scratch every time. ROS has high level commands for anything a robot can do and becomes increasingly useful as robot complexity grows. It starts its operation in a robot first by obtaining information about the robot's characteristics, such as the length and movement limits of limbs, servos involved, sensors and then it will communicate this data to higher level algorithms. Apart from running physical robots, as discussed under the term robotics simulators, designing robots in virtual environments first has many advantages. ROS also has a simulator called TurtleBot, which allows to simulate and test robots without having a physical robot, only in a virtual environment. This alone can speed up robot development process remotely. ROS is written in C++ and Phyton. When ROS first came out, it was C++ only.

There is also H-ROS, which stands for Hardware - Robot Operating System, which was introduced in 2016 and supported by several more companies. Each piece runs ROS 2.0 on its own has its own ROS nodes and topics. Building robots is about putting together different H-ROS components that can come from different manufacturers but still inter-operate and exchangeable thanks to the standard hardware interfaces defined within H-ROS. This also allows developers to easily upgrade their robots with hardware from different manufacturers and add new features in seconds. With H-ROS, building robots may only mean about placing H-ROS-compatible hardware components together to build new robot configurations, and this will enable a greater number of people to be able to construct robots, and not only people with technical skills. The blocks that make up the system fall into 5 categories, which are, sensing, actuation, communication, cognition and hybrid components. H-ROS was initially funded by the US Defense Advanced Research Projects Agency (DARPA) through the Robotics Fast Track program in 2016 and developed by Erle Robotics of Spain. The platform has already been tested by several international manufacturers who have built robots out of this technology. H-ROS is also able to use the ROS simulator TurtleBot, which was mentioned in previous paragraph. See H-ROS website at https://H-ROS.com/ for more details on this.

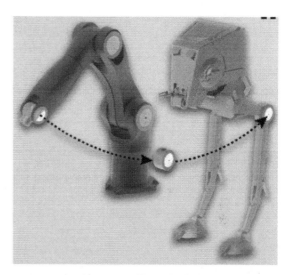
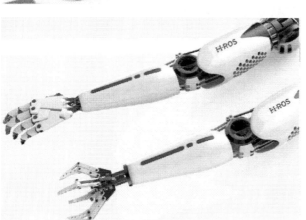
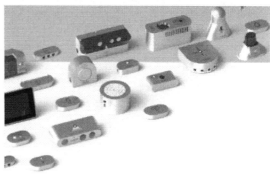

Photo Credits: Erle Robotics - https://www.H-ROS.com/, www.erlerobotics.com

Industrial Robot Programming

Many industrial robot manufacturers write their own programs to control robots, but there are also a great deal of firms, specifically dealing with industrial robot control software and programming, which can be applied to robots of different brands. Some important features of an industrial robot software include:

- Applicability to different robot brands
- Force control in 6D, which includes movement and rotation along all 3 axes
- Dealing with error, geometrical differences, and variable situations
- Ability to be integrated with different sensors
- Human - Robot shared, or, total autonomous operation
- Ease of programming
- Option to be programmed manually
- Tolerance to different lighting conditions
- Safety features

CHAPTER 12

VARIOUS ROBOTS

Other than the robot types that were presented in this book, there are many other robots, that do not fit into any of those categories we mentioned before. They vary greatly in their shape, size, what they do and their purpose. Out of so many robots which are all very different from each other in every way, we would like to mention a few:

Sub 1 Reloaded - Rubik Cube Solver:
Solving Rubik cube is an impressive feat for humans, but robots can now do it in fraction of a second. In 2016, a German company called Infineon showed its Rubik Cube solver robot, called Sub 1 Reloaded, which was able to solve the cube in 0.637 seconds. The company used a microcontroller that would be used in autonomous driving systems, in order to achieve the record.

Ballbot - Planetary Mission Robot:
Actually the term "ballbot" is a type of robot that was presented under *Robotics Terms and Concepts* chapter but the robot called Ball Bot made by NASA, does not fit in that category at all, as it has only a name resemblance and completely different than a usual ballbot. This robot was designed for planetary missions. The structure is specially designed for high impact resistance which helps to reduce the cost and reliability. The structure, which is called "tensegrity", which is based on tension and compression members, is very useful in climbing slopes and handling rough terrain, in addition to high impact resistance. This is a very lightweight robot, weighing few kilograms, and can be conveniently packed during launch, and would reliably separate and unpack at the destination. This is a robot which is still in development. Source: https://www.nasa.gov/content/super-ball-bot

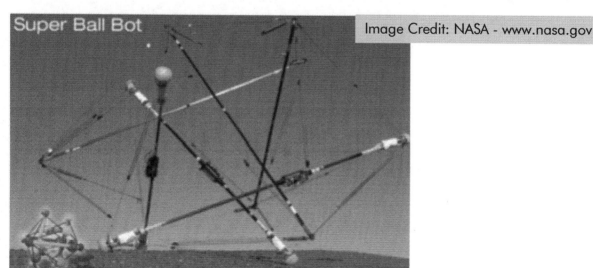

Super Ball Bot

Image Credit: NASA - www.nasa.gov

Curiosity Rover - Planetary Mission Robot:

NASA's Curiosity Rover was launched on November 2011 and landed on Mars August 2012. Its mission is to investigate Mars surface for past and present conditions and to gather data, in order to see if the planet once sustained life or may be able to sustain life, even at the microbe level. The robot, which weighs 900 kg (2000 lbs), was first getting all orders from earth but recently the engineers at NASA installed an AI software to the rover's main computer and now it is able to use its own intelligence to compliment human decisions, to decide which locations to explore, to avoid loosing time due signal delays because of the long distance between Earth and Mars. The body of the robot is high of the ground, so that it will not stuck between any rocks. Its battery sends the excess heat to its computer, so that it will be kept warm under extremely cold temperatures on Mars. The robot can drill rocks with its robot arm, which has a reach of 2.2 meters (7 feet) and then it can grind the samples to powder, to be analyzed inside its body which contains scientific instruments. The spinning of the wheels are controlled by a special software, to get the maximum traction and also to minimize wearing of tires, by reducing pressure from the rocks as much as possible in any situation. At one point, when the robot's main computer broke down, which was called the pilot, the second computer, which was called the co-pilot took over, and it still serves as the controlling computer of the robot. Engineers from Earth, were still able to fix the main computer remotely, through commands sent via radio signals. The robot has covered more than 15 km distance in 5 years on Mars surface, in a 96 mile wide crater called Gale crater. Curiosity Rover was manufactured by Boeing, Lockheed Martin and Jet Propulsion Laboratory. Source: https://www.jpl.nasa.gov, https://mars.nasa.gov/msl/mission/rover/

Images Courtesy NASA/JPL-Caltech

Robotic Animals

Imitating nature is an important part of robotics. Analyzing how animals naturally move and applying these principles to robots help us to create robots that can move better and perform much more efficiently. For this reason, a lot of firms work on creating robotic animals, in order to create better performing robots, which can be useful for a lot of different tasks. Below we introduce just a few examples, but needless to say, there are many more types and examples, that were either built or being worked on.

Robot Bird:

The robot bird which practically flies like a real bird by flapping its wings, is used to chase birds away from the airports was made by a Dutch company called Clear Flight Solutions. The real birds think that the robot, which is made in the shape and size of a falcon and is hardly distinguishable from a real falcon from outside, is their natural enemy and try to avoid it. https://clearflightsolutions.com/

Another robot bird, made by the German company called Festo, is designed in the shape of a Seagull, and also has impressive flight capabilities by flapping its wings and use wind uplift. https://www.festo.com/group/en/cms/10238.htm

Robot Butterfly:

The robot butterfly made by the German company Festo is hard to distinguish from real. It can either fly autonomously using GPS, or in swarms, controlled by a central computer, which watches the swarm with camera. It was made by using the minimum materials and as lightweight as possible. https://www.festo.com/group/en/cms/10216.htm

Robot Dog:

Spotmini: This is a four legged robot, created by Boston Dynamics of USA, designed for home use, which also includes a robotic arm and gripper, in order to be able to manipulate things. The robot can walk on any surface, including climbing stairs easily, walk by crouching under openings which are lower than its normal height or get up after an unexpected fall. Its robotic 5 DOF arm and gripper also seem to be very skillful to perform a variety of tasks, such as correctly identifying a glass and an empty can of soda, and placing them easily in the dishwasher and the trash can respectively, or bringing a can of coke to a human sitting on a couch. It is also able to maintain the position of the end of its gripper, even if it moves or rotates its whole body in any direction. The robot weighs 30 kg, measures 84 cm high, and can carry a payload of 14 kg. It is able to work very silently on a single battery charge, for up to 90 minutes. It has sensors for both manipulation and navigation, such as cameras, IMU, position and force sensors. https://www.bostondynamics.com/spot-mini

Robot Salamander:

A robot named Pleurobot has the shape of a salamander skeleton, which was developed by Swiss Federal Institute of Technology at Lausanne. It was 3D printed, can walk, swim and made to understand amphibian locomotion. It was constructed by tracking many points on a real salamanders body while in motion. Its electrical circuitry mimics a nervous system and has 27 degrees of freedom. https://biorob.epfl.ch/pleurobot

Photo Credit: Biorobotics Laboratory, EPFL, Switzerland
http://biorob.epfl.ch/pleurobot

CHAPTER 13

ROBOT & ROBOT RELATED COMPANIES LIST

The companies below are relevant to the scope of this book as we determined and were able to identify. No payments was received from any company to be included in this list or anywhere in this book. Author does not have association with any of the companies mentioned or listed in this book.

Names and brand names: For each company we wrote either as:

Trademark name - company name who owns that trademark - web address - short description

OR, if the trademark and company name are the same, we only wrote as:

Company / Trademark name - web address - short description

Unless there are errors, the brand and/or company names are written exactly the same way we have seen on company websites, including all capitalization and lower case letters, to respect trademarks.

Descriptions: The descriptions of companies below are considerably shortened several word summaries. They may not be complete or very accurate. We wrote these to the best of our ability from robotics point of view, to help readers see at once and in just several words what each company does. Therefore, if a company description seems even remotely relevant in scope, to what the reader is looking for, we advise the reader to visit the company website, and verify scope.

What is included and excluded: Very large companies, that operate in many fields, are generally not included, unless their robotics work is well known. Multinational robotics companies, are generally listed only under their country of world headquarters. Parts producers, robotics service providers are not included with some exceptions. Seller companies or online stores are also excluded, except the well known, and established large volume stores that we were aware of, which we thought might be useful to our readers. Therefore, this is generally a list of end product makers, who actively create useful end products or solutions, as robots or things very directly related to robots. Bottom line is that, we wanted to make a current and useful list of worldwide robot and directly robot related companies here, which would be useful to readers from all levels and perspectives.

Notes: The company list below may not be copied. All company and product names mentioned in this book and listed below are trademarks that belong to their respective owners. Best efforts were spent to respect trademarks and write spellings and descriptions correct and any possible errors in the list below or within the prior text of the book are unintentional. If you feel there should be a correction to what we wrote about your firm, or want your firm removed, please contact us at **editor@roboticmagazine.com** and we will do it in the next version of this book. You can also ask us to be included if you think your company should also be in here. This is a free listing but we can only include relevant firms, therefore we cannot guarantee inclusion. Note that links or facts may change after the publication date of the book.

Australia

Apex Automation & Robotics Pty Ltd - http://www.apexautomation.com.au/ - robotic systems and automation

Aubot Pty Ltd - http://aubot.com/ - telepresence robot and robot arm manufacturer

Fast Brick Robotics Ltd - http://fbr.com.au/ - developer of brick laying robots

FREELANCE ROBOTICS - http://www.freelancerobotics.com.au - robotics research, development and custom integration company

Impact Robotics - http://www.impactrobotics.com.au/ - robot integration company for industrial process improvement

Ocular Robotics Limited - http://www.ocularrobotics.com/ - manufacturer of LIDAR, robotic vision systems

Mexx Engineering - http://www.mexx.com.au/ - robotics automation, welding cells, conveyor and vision systems

Robot Technologies Systems Australia Pty Ltd - http://www.robottechnologies.com.au/ - robot systems integrator

SCOTT Technology Limited - https://www.scottautomation.com/ - automation and robotic solutions for industrial application

Silvertone Electronics - http://www.silvertone.com.au- UAV manufacturer

Textron Systems Australia Pty Ltd. - http://www.textronsystems.com/ - unmanned systems manufacturer for aerospace industry

V-Tol Aerospace Pty Limited - http://v-tol.com/ - UAS manufacturer

Zodiac Group Australia Pty Ltd - https://www.zodiac.com.au/ - manufacturer of pool cleaning robots

Austria

AIRBORNE ROBOTICS GMBH -http://www.airborne-robotics.com/ - UAV manufacturer

B & R - https://www.br-automation.com/ - automation parts and products manufacturer

Dynamic Perspective GmbH - http://dynamicperspective.com/home/ - UAV manufacturer

Flitework - http://flitework.at/ - RC plane manufacturer

igm Robotersysteme AG - http://www.igm-group.com/en - welding robot manufacturer

Shiebel Aircraft GmbH - http://www.schiebel.net/ - UAV manufacturer

Belarus

Indela - http://www.indelauav.com/ - drone manufacturer

Belgium

Altigator - http://altigator.com/en/ - drone manufacturer, seller, drone services

Flexible Robotic Solutions - http://www.frsrobotics.com/ - developer of sensor based robot control software and automation

Brasil

XBot - http://www.xbot.com.br/ - developer of educational robots

Aerofoundry - http://www.aerofoundry.com/ - drone manufacturer

FT Sistemas - http://flighttech.com.br/ - drone manufacturer

Canada

4Front Robotics - http://www.4frontrobotics.com/ - developer of UAV, UGV, robot hands and navigational systems

Aeromao Inc. - http://www.aeromao.com/ - drone manufacturer, seller, drone services

Aeryon Labs Inc. - http://aeryon.com/ - drone manufacturer

ARGO - http://www.argorobotics.com/ - manufacturer of extreme terrain UGVs

BC Robotics Inc. - http://www.bc-robotics.com/ - manufacturer of electronic kits and components

Bionik Laboratories Corp. - http://www.bioniklabs.com/ - bioengineered rehabilitation solutions

BKIN Technologies - http://www.bkintechnologies.com/ - developer or medical robots

Canada Robotix - http://www.canadarobotix.com/ - robot parts and components seller

Challis Helicopters Inc. - http://www.challis-heliplane.com - drone manufacturer

Clearpath Robotics Inc. - https://www.clearpathrobotics.com/ - maker of autonomous mobile robots for land, sea or air, industrial robot material transporters, various robotics accessories

GaitTronics Inc. - http://gaittronics.com/ - developer of physical rehabilitation robots

Hypertherm Inc. - http://www.robotmaster.com/en/ - developer of software for industrial robots

ING Robotic Aviation - http://ingrobotic.com/ - drone manufacturer, drone services

Kinova Robotics - http://www.kinovarobotics.com/ - manufacturer of robots that help people with disabilities, domestic task helper robots

OCTOPUZ Inc. - http://octopuz.com/ - industrial robot programming software developer

Opus Automation - http://www.opusautomation.com/ - robotic automation solutions developer

RobotShop Inc. – http://www.robotshop.com - hobby and service robots & parts seller, service provider

Robotiq - http://www.robotiq.com - developer of flexible robotic grippers

Stratus Aeronautics - http://www.stratusaeronautics.com/ - drone manufacturer

Titan Medical Inc. - http://www.titanmedicalinc.com/ - surgical robots manufacturer

Xaxxon Technologies - http://www.xaxxon.com/ - developer of autonomous mobile robots

WowWee Group Limited - http://wowwee.com/ - developer of robotic toys

China

AmTidy - http://www.amtidy.com/ - robotic vacuum cleaner manufacturer

AvatarMind - http://www.avatarmind.com/ - developer of humanoid companion robots

Cheerson - Guangdong Cheerson Hobby Technology Co., Ltd. - http://www.cheersonhobby.com/ - drone manufacturer

Dexta Robotics - http://www.dextarobotics.com/ - developer of robotic hands

Diatone Innovations Co.,Ltd. - http://www.diatone.hk - drone manufacturer

DJI - http://www.dji.com/ - drone manufacturer

Dualsky - Shanghai Dualsky Models Co., Ltd. - http://www.dualsky.com/ - drone parts manufacturer

ECOVACS ROBOTICS - http://global.ecovacs.com/ - manufacturer of robotic vacuum cleaners

EHANG - http://www.ehang.com/ - drone manufacturer

Emlid Ltd - http://www.emlid.com - drone manufacturer, seller

Feishen Group Company - http://www.fs-racingart.com/uav.php - drone manufacturer

FLYPRO - Shenzhen FLYPRO Aerospace Tech Co., Ltd. - http://www.flypro.com/en - drone manufacturer

Hanson Robotics Limited - http://www.hansonrobotics.com/ - manufacturer of humanoid robots

Han's Motor - Shenzhen Han's Motor S&T Co., Ltd. - http://www.hansmotor.com/en/industrialrobots.html - manufacturer of industrial robots and parts

HEXAR SYSTEMS - http://hexarsystems.com/new/ - manufacturer of walking assistance robots

Hubsan - http://www.hubsan.com/ - drone manufacturer

Idea Fly - Shenzhen Idea-Fly Technology Co., Ltd - http://www.idea-fly.com/ - drone manufacturer

IKV Robot Nanchang Co., Ltd. - http://www.ikvrobot.com/ - industrial robot manufacturer and automation solutions provider

INMOTION TECHNOLOGIES CO., LTD - http://www.imscv.com - self balancing transporter robot manufacturer

Keweitai - Shenzen Keweitai Enterprise Development Co. Ltd. - http://en.keweitai.com/ - drone manufacturer

Keyirobot - http://www.keyirobot.com/ - educational modular robot developer

Makeblock Co., Ltd. - http://store.makeblock.com/ - open source robot kits developer

Makerfire - Shenzhen Makerfire Technology Co. Ltd. - http://www.makerfire.com/ - drone manufacturer

Midea Group - http://www.midea.com/global/ - manufacturer of a wide variety of electronics equipment including robotics

MEIJIAXIN TOYS CO.,LIMITED - http://www.mjxtoys.com/ - drone manufacturer

Nine Eagles - Shangai Nine Eagles Electronic Technology Co., Ltd. - http://www.nineeagle.com/ - drone manufacturer

NXROBO - http://www.nxrobo.com/ - developer of personalized family robots

Powervision - http://powervision.me/uk/ - manufacturer of drones, including underwater drones

Radiolink Electronic Limited. - http://www.radiolink.com.cn/doce/index.html - Transmitter, receiver manufacturer

RC Logger - CEI Conrad Electronic Intl. (HK) Ltd. - http://www.rclogger.com/ - drone manufacturer

Sanbot - Qihan Technology Co., Ltd. - http://en.sanbot.com/index.html - developer of robotics, artificial intelligence, and video analysis technologies

Servosila - https://www.servosila.com/en/ - developer of mobile robots, robotic arms, servo drives, robotic control systems, software packages

Shenzhen AEE Technology CO., LTD - http://www.aee.com - drone manufacturer

SkyRC Technology Co., Ltd.- http://www.skyrc.com/ - drone parts manufacturer

Skyartec - R/C Model Fun., CO LTD - http://www.skyartec.com/ - drone manufacturer

SYMA - Guangdong Syma Model Aircraft Industrial Co., Ltd. - http://www.symatoys.com/ - drone manufacturer

T-MOTOR - http://www.rctigermotor.com/ - drone motor, ESC, propeller manufacturer

TAROT RC - WENZHOU TAROT AVIATION TECHNOLOGY CO.,LTD- http://www.tarotrc.com - drone and drone parts manufacturer

UBTECH - http://www.ubtrobot.com/ - manufacturer of educational robots and kits

Walkera - Guangzhou Walkera Technology CO., LTD - http://www.walkera.com/ - drone manufacturer

Worx America - Resort Savers, Inc. - - http://resortsaversinc.com/worx-america/ - automated engineering solutions for industrial, environmental, energy sector applications

X-Robot - Shanghai New Century Robot Co., Ltd - http://www.x-robot.net - manufacturer of self mobility robots

XY Aviation - Xingyu Zhuhai Aviation Technology Co., Ltd. - http://www.xy-aviation.com/ - drone manufacturer

Yuneec International Co. Ltd. - http://www.yuneec.com/ - drone and drone accessories manufacturer

Croatia
DOK-ING doo - http://www.dok-ing.hr/ - manufacturer of mine clearance robots

Czech Republic

AXi Model Motors - MODEL MOTORS S. R. O. - https://www.modelmotors.cz/ - drone motor manufacturer

JETI model s.r.o. - http://www.jetimodel.com/en/ - electronic parts manufacturer for drones

MS Composit - Jan Hess-Modelsport - http://www.mscomposit.info/ - drone manufacturer

Denmark
Blue Ocean Robotics - http://www.blue-ocean-robotics.com/ - developer of various industrial and commercial robots

Blue Work Force - http://blueworkforce.com/ - developer of 3D printing and handling robots

Danish Aviation Systems - http://www.danishaviationsystems.dk/ - drone manufacturer

LEGO - https://www.lego.com/ - manufacturer of learning and development toys including robot kits

Mobile Industrial Robots ApS - http://mobile-industrial-robots.com/en/ - mobile robot manufacturer for logistics and internal transportation

Sky-Watch A/S - http://sky-watch.dk/ - drone manufacturer

Universal Robots A/S - https://www.universal-robots.com/ - manufacturer of collaborative robotic arms (cobots)

Estonia

Starship Technologies - https://www.starship.xyz/ - manufacturer of autonomous delivery robots

THREOD SYSTEMS - http://www.threod.com/ - drone manufacturer

Finland

AeroTekniikka UAV Oy - http://www.aerotekniikka.fi/ - drone manufacturer

Arctic Robotics Oy - https://www.arcticrobotics.com/ - indoor drone manufacturer

Robbo Europe - http://robbo.world/products/ - educational robot kits developer

Robot Power Finland Oy - http://www.robotpower.fi - robotic automation solutions provider

VideoDrone Finland Oy - http://www.videodrone.fi/ - drone manufacturer

ZenRobotics Ltd. - http://zenrobotics.com/ - developer of waste separation robots

France

Aldebaran Robotics (Owned by Softbank Robotics) - https://www.ald.softbankrobotics.com/en - humanoid robot manufacturer

BALYO- http://www.balyo.com/en - developer of automated guided vehicles

BLUE FROG ROBOTICS - http://www.bluefrogrobotics.com/ - developer of companion robots

Delair-Tech - http://www.delair-tech.com/en/home/ - drone manufacturer

DIBOTICS - http://www.dibotics.com/ - developer of navigation and 3D sensing algorithms

E.ZICOM - http://www.e-zicom.com - manufacturer of robotic vacuum cleaners

Flying Robots - http://www.flying-robots.com/en/ - drone manufacturer

HELIPSE - http://www.helipse.com/ - drone manufacturer

LIRMM - http://www.lirmm.fr/lirmm_eng - informatics, robotics and microelectronics research

MIP Robotics - http://www.mip-robotics.com - manufacturer of collaborative, industrial robots

MOBILITE SERVICE - http://www.mobilite-service.fr - developer of transport robots

Novadem - http://www.novadem.com/ - drone manufacturer

RB3D - http://www.rb3d.com/en/ - developer of robotic manipulation and handling tools, exoskeletons

Reeti - Robopec - http://www.reeti.fr - companion robot platform developer

Robopec - http://www.robopec.com - software and robotic architecture systems developer

ROBOSOFT - http://robosoft.com/ - developer of autonomous vehicles for various industries

Robotswim - http://www.robotswim.com - designer of underwater robots

SBG Systems - http://www.sbg-systems.com/ - inertial sensor manufacturer

Sepro Group - http://www.sepro-group.com/ - industrial robot manufacturer

Softbank Robotics Europe - See Aldebaran Robotics

SpirOps - http://www.spirops.com/ - artificial intelligence research

SURVEY Copter - http://www.survey-copter.com/ - drone manufacturer

Wandercraft - http://www.wandercraft.eu/ - robotic exoskeleton manufacturer

Xamen Technologies - http://www.xamen.fr/index.php/fr/ - UAV manufacturer

Germany

Aero-naut Modellbau GmbH- http://www.aero-naut.de/en/home/ - model RC manufacturer

Aibotix GmbH - https://www.aibotix.com/en/ - UAV manufacturer

AirRobot GmbH & Co. KG - http://www.airrobot.com/ - UAV manufacturer

ArtiMinds Robotics GmbH - https://www.artiminds.com/ - developer of robotic software for sensor adaptive motion

ASCENDING TECHNOLOGIES - Intel Deutschland GmbH - http://www.asctec.de/ - UAV manufacturer

BAVARIANDEMON - CAPTRON Electronic GmbH - http://www.bavariandemon.com/en/ - electronic parts manufacturer for drones

birdpilot GmbH- http://www.birdpilot.com/en/home - multicopter drone manufacturer

Braeckman Modellbau - http://www.braeckman.de/ - drone manufacturer

CAPTRON Electronic GmbH - http://www.captron.de/en/ - capacitive and optical sensor manufacturer

Care-O-bot 4 - http://www.care-o-bot-4.de/ - manufacturer of mobile humanoid robot for domestic use

Commonplace Robotics GmbH - http://www.cpr-robots.com/ - developer of robots for education, research and industry

D-Power - http://www.d-power-modellbau.com/ - drone manufacturer and RC products seller

EASY-ROB - http://www.easy-rob.com/easy-rob/ - software developer for manufacturing plant workcells

Eureka - Roboris-Deutschland GmbH - http://roboris-deutschland.de/ - robotics simulation software developer

Festo Vertrieb GmbH & Co. KG - https://www.festo.com - biologically inspired robots, gripping systems, pneumatic systems, various industrial robotics applications

fischertechnik GmbH - http://www.fischertechnik.de - educational kits manufacturer

fsk engineering GmbH - http://www.fsk-engineering.de - developer of robot cell programs and simulators

German Robot - http://www.german-robot.com/ - open source humanoid robot developer

Graupner/SJ GmbH - http://www.graupner.de - RC products manufacturer, seller

HBi Robotics - http://www.hbi-robotics.de/ - automation solutions provider

Infineon Technologies AG - https://www.infineon.com/ - semiconductor solutions and a variety of microelectronics manufacturer

Innok Robotics GmbH - http://www.innok-robotics.de - modular outdoor robot manufacturer iRC-Electronic GmbH- http://shop.rc-electronic.com/ - RC products and accessories manufacturer and seller

KONTRONIK Drives - SOBEK Drives GmbH - http://www.kontronik.com - motor and electronic components manufacturer for drones

KUKA AG - https://www.kuka.com/ - industrial robot manufacturer

LF-Technik GmbH - http://www.lf-technik.de/ - drone manufacturer

Magazino GmbH- http://www.magazino.eu - manufacturer of warehouse robots

MetraLabs GmbH - http://www.metralabs.com/en/ - manufacturer of telepresence, retail robots and mobile robot platforms

Microdrones GmbH - https://www.microdrones.com/en/home/ - drone manufacturer, drone services

MIKADO Model Helicopters GmbH - http://shop.mikado-heli.de/ - model helicopters and accessories manufacturer, seller

MikoKopter - HiSystems GmbH - http://www.mikrokopter.de/en/home - drone manufacturer, drone services

Mikrotron GmbH - http://www.mikrotron.de/en.html - high speed visual recognition systems

Miniprop GmbH - http://www.miniprop.com/ - drone manufacturer

MRK-Systeme GmbH- http://www.mrk-systeme.de/ - industrial robot manufacturer

Multiplex - http://www.multiplex-rc.de/ - RC products and accessories manufacturer, seller

Neobotix GmbH - http://www.neobotix-robots.com/ - manufacturer of mobile robots and robot systems, manipulators

Nerian Vision Technologies - https://nerian.com/ - visual depth sensor developer

Otto Bock HealthCare Deutschland GmbH - http://www.ottobock.de/ - developer of prosthetic robotics

Plettenberg Elektromotoren- http://www.plettenberg-motoren.net/ - motor drone manufacturer

rOsewhite Multicopters - http://www.rosewhite.de/ - drone manufacturer

Rowa Technologies - http://rowa.de/en/pharmacist - pharmaceutical robot manufacturer

SCHUNK GmbH & Co. KG - https://schunk.com - manufacturer of gripping and various systems for industrial robots

service-drone.de GmbH - https://www.service-drone.com/en/ - drone manufacturer, drone services

SOBEK Drives GmbH - http://sobek-drives.de/ - motor and electronic components manufacturer for drones

Tinkerbots - Kinematics GmbH - https://www.tinkerbots.com/ - robot toys manufacturer

VESCON - http://www.vescon.com - automation solutions provider

xamla - PROVISIO GmbH - http://xamla.com/en/ - developer of open source assistive robotic solutions

Zimmer Group - http://www.zimmer-group.de - components manufacturer for industrial robots

Iceland

Ossur - https://www.ossur.com - robotic prosthetics developer

India

ASIMOV ROBOTICS PVT LTD- http://www.asimovrobotics.com/ - maker of stationary or mobile robot manipulator arms, humanoid robots, simulation, machine vision, virtual reality consultancy

GREYORANGE - http://www.greyorange.com/ - developer of warehouse robotics solutions

Kadet Defence Systems (P) Limited - http://www.kadet-uav.com/ - drone manufacturer

Milagrow Business & Knowledge Solutions (Pvt.) Limited. - http://milagrowhumantech.com/ - developer of consumer robots

Nex Robotics - http://www.nex-robotics.com/ - robotics platform design and manufacturing

Robosoft Systems - http://www.robosoftsystems.co.in/ - duct inspection robots, robotic control systems manufacturer, various robots and parts seller

Sastra Robotics India Pvt Ltd - http://www.sastrarobotics.com/ - developer of industrial manipulators and telepresence robots

Indonesia

Aero Terrascan - PT. Aero Terra Indonesia- http://www.aeroterrascan.com/ - drone manufacturer, drone services

Ireland

EiraTech robotics - http://www.eiratech.com/ - warehouse robotics solutions developer

Reamda Ltd - http://reamda.com/ - manufacturer of mobile robots with manipulators

Robotics & Drives - http://www.roboticsanddrives.ie/ - robotics automation solutions

Tricon Automation Ireland Ltd.- http://www.tricon.ie/ - developer of automation and robotics solutions

Israel

BlueBird Aero Systems - http://www.bluebird-uav.com/ - drone manufacturer

Cogniteam - http://cogniteam.com - maker of mapping mobile robot, autonomous control and interactive simulation developer

CONTROP Precision Technologies Ltd. - http://www.controp.com/ - electro-optical and precision motion control systems manufacturer

Innocon - http://www.innoconltd.com/ - UAS manufacturer

Israel Aerospace Industries Ltd. - http://www.iai.co.il - manufacturer of unmanned systems and various military equipment and vehicles for defense industry

Mazor Robotics - https://www.mazorrobotics.com/ - manufacturer of medical surgery robots

RoboTiCan Ltd. - http://www.robotican.net/ - mobile robotics and manipulators developer

Steadicopter Ltd. - http://www.steadicopter.com/ - drone manufacturer

Italy

Aermatica3D srl- http://www.aermatica.com/ - drone systems integrator

CMA ROBOTICS SPA ITALY - http://www.cmarobot.it/index.php - industrial robot manufacturer

Comau S.p.A. - http://www.comau.com/IT - robotics automation solutions

Gimatic S.r.l. - http://www.gimatic.com/en - manufacturer of pneumatic and electric grippers for industrial automation

IUVO S.r.l. - http://www.iuvo.company/home.html - wearable robotic devices developer

Kinetek - Wearable Robotics srl - http://www.wearable-robotics.com/kinetek/ - robotic exoskeleton manufacturer

MAVTech s.r.l. - http://www.mavtech.eu/ - drone manufacturer

MSHeli Srl - http://www.msheli.com/ - drone manufacturer

Prensilia Srl - http://www.prensilia.com/ - manufacturer of anthropomorphic, under-actuated robotic hands

SAB HELI DIVISION - http://www.sabitaly.it/ - manufacturer and seller of RC products and parts

T&D Robotics S.r.l. - http://www.tdrobotics.com/ - manufacturer of industrial robots

VisLab srl - http://vislab.it/ - developer of machine vision algorithms and intelligent systems for automotive industry

Japan

CYBERDYNE, INC. - http://www.cyberdyne.jp/english/ - developer of robotic exoskeletons

Drone Express Department Strawberry Cones Co., Ltd. - http://www.droneex.net/en/ - drone manufacturer

EPSON Robots - http://robots.epson.com/ - industrial robots manufacturer

FANUC CORPORATION - http://www.fanuc.com/ - industrial robots manufacturer

Flower Robotics, Inc. - http://www.flower-robotics.com/ - consumer robots manufacturer

HiBot - https://www.hibot.co.jp/ - pipeline Inspection robot developer

KAWADA ROBOTICS CORPORATION - https://www.kawadarobot.co.jp/index_en.html - manufacturer of humanoid and industrial robots

MUJIN INC. - http://mujin.co.jp/en/ - industrial robots developer

NACHI-FUJIKOSHI CORP. - http://www.nachi-fujikoshi.co.jp/eng/index.html - industrial equipment and robot manufacturer

ROBO GARAGE CO., LTD. - http://www.robo-garage.com/top.html - humanoid toy robots manufacturer

SoftBank Robotics Corp. - http://www.softbank.jp/en/robot/ - developer of humanoid robots and various electronic equipment

YASKAWA ELECTRIC CORPORATION - https://www.yaskawa.co.jp/en/- ac drives, motion control products, servos, industrial robots manufacturer

Yukai Engineering Inc. - http://www.ux-xu.com/ - developer of personal robots

ZMP INC. - https://www.zmp.co.jp - autonomous technologies developer

Latvia

Drone Technology - http://www.dronetechnology.eu/ - drone manufacturer

RobotNest - http://www.robot-nest.com/ - robotics projects developer

Mexico

ACS Automatizacion - http://www.acssa.com/ - robotics automation solutions provider

Advance Industrial Robotics - http://www.adrobotics.com.mx/ - robotics automation solutions provider

Automatische Technik - http://www.atechnik.com.mx/ - industrial robots manufacturer

briko - https://brikorobotics.com/ - educational robot kit manufacturer

Industrial Robotics Solutions México, S.A. DE C.V. - http://www.industrialrobotics.com.mx - robotics automation solutions provider

Robotica Industrial Proyectos México S.A. de C.V - http://www.roboticaindustrial.com.mx/ - - robotics automation solutions provider

Netherlands

AceCore Technologies - http://www.acecoretechnologies.com/ - drone manufacturer

AERIALTRONICS - http://www.aerialtronics.com/ - drone manufacturer

Aerobotica - http://aerobotica.nl/ - drone manufacturer

Birds-Eye-View - https://www.birds-eye-view.nl/ - drone and RC products parts manufacturer, seller

Clear Flight Solutions - http://www.clearflightsolutions.com/ - drone manufacturer

Delft Dynamics - http://www.delftdynamics.nl/index.php/en/ - drone manufacturer

Exact Dynamics - http://www.exactdynamics.nl - developer of robotic manipulators

Focal Meditech - http://www.focalmeditech.nl/en - developer of medical aid robots

HIGH EYE UNMANNED AVIATION- http://www.higheye.nl/company/ - drone manufacturer

Polymac Robotics b.v. - http://www.polymac.nl/en/ - industrial robots and automation solutions developer

Smart Robot Solutions - http://smartrobot.solutions/ - telepresence robot manufacturer

New Zeland

Aeronavics Ltd. - http://aeronavics.com/ - drone manufacturer

Rex Bionics Ltd - http://www.rexbionics.com/ - manufacturer of robotic mobility devices

ROBOTICS PLUS - http://www.roboticsplus.co.nz/ - agricultural robots developer

SKYCAM UAV - http://www.kahunet.co.nz/ - drone manufacturer, drone services

Norway

Blueye Robotics - https://gust.com/companies/blueeye - manufacturer of underwater drones

Maritime Robotics - http://www.maritimerobotics.com/ - manufacturer of unmanned sea vehicles

Proxy Dynamics AS - http://www.proxdynamics.com/home - drone manufacturer

Robomop International AS - http://robomop.com/ - robotic floor duster manufacturer

Robot Aviation - http://www.robotaviation.com/ - drone manufacturer

Pakistan

Integrated Dynamics - http://www.idaerospace.com/ - drone manufacturer

SATUMA - http://www.satuma.com.pk/ - drone manufacturer

Portugal

Motofil Robotics - http://motofil.pt/index.php/en/home/ - industrial robot manufacturer

TURFLYNX Iberia - http://www.turflynx.com/ - driverless lawn mower manufacturer

UAVision Lda - http://www.uavision.com/ - drone manufacturer

Romania

Autonomous Flight Technologies - http://www.aft.ro/ - drone manufacturer

AXOSUITS SRL - http://axosuits.com/ - medical exoskeleton manufacturer

Russia

ANDROID TECHNICS - http://en.npo-at.com/ - developer of anthropomorphic robotic systems

EXOATLET - http://exoatlet.ru - exoskeleton manufacturer

Rbot - http://rbot.com/ - developer of consumer robots

RoboCV - http://robocv.com/ - developer of autopilots for warehouse vehicles

VisionLabs - http://www.visionlabs.ru/en/ - developer of visual recognition software

Singapore

CtrlWorks Pte Ltd - http://www.ctrlworks.com/ - developer of wheeled robots

Infinium Robotics - http://www.infiniumrobotics.com/ - drone manufacturer

O'Qualia - http://www.oqualia.com/ - drone manufacturer, drone services

Rotimatic - Zimplistic Pte Ltd - https://rotimatic.com/ - manufacturer of flatbread making robots

SESTO Robotics - http://sestorobotics.com/ - AGV manufacturer

Slovenia

C-Astral d.o.o. - http://www.c-astral.com/ - drone manufacturer

South Africa

Aerial Monitoring Solutions - http://www.ams.za.com/ - drone manufacturer

CATUAV - http://www.catuav.com/ - drone manufacturer

Embention - http://www.embention.com/en/home.htm - developer of unmanned vehicle components and complete autonomous systems

Micro Robotics - https://www.robotics.org.za/ - robotics and electronics parts seller

Robotic Innovations - http://www.roboticinnovations.co.za/ - industrial robot solutions provider

Ryonic Robotics - http://www.ryonic.io/ - manufacturer of pipeline and hull cleaning robots, mine explorer, collaborative industrial robots

South Korea

FUTUREROBOT Co., Ltd. - http://www.futurerobot.co.kr - manufacturer of smart home robots

Gryphon Dynamics - http://gryphondynamics.co.kr/ - drone manufacturer

KOBUKI - Yujin Robot - http://kobuki.yujinrobot.com/ - developer of mobile robotics platforms

Naran Inc. - https://prota.info/microbot/ - developer of smart hubs, micro robots for automating existing appliances

NEOFECT - http://www.neofect.com/en/ - wearable robotic hand manufacturer

robo3 - http://www.robo3.com/ - manufacturer of personal mobility and transporter robots

Robolife Inc. - http://www.robo-life.co.kr/ - humanoid and educational robot kits manufacturer

roborobo Co., Ltd. - http://www.roborobo.co.kr/ - educational robot kits manufacturer

Robobuilder - http://www.robobuilder.net/ - developer of educational humanoid and robot kits

ROBOTIS - http://en.robotis.com/index/ - manufacturer of robotic platforms, educational kits, humanoid robots, software, actuators

Robotron B/D - http://www.robotron.co.kr/ - hobby robot kits manufacturer

SRC - http://src.koreasme.com/ - humanoid robot kit manufacturer

UMAC Air - http://www.umacair.com/ - drone manufacturer

WONIK ROBOTICS - http://www.simlab.co.kr/ - software and hardware solutions, including robotic hand, software and applications for various manipulators, mobile robots developer

YUJIN ROBOT CO., LTD - http://en.yujinrobot.com/ - developer of service robots

Spain

Alpha Unmanned Systems - http://www.alphaunmannedsystems.com/ - drone manufacturer

Aquatron Robotic Technolgy ltd. - http://www.aquatron.co.il/ - robotic pool cleaner manufacturer

Erle Robotics S.L. - http://erlerobotics.com/blog/home-creative/ - hobby robots, drone autopilots, multirotor drones manufacturer, robotics project management, ROS developer, system design

Idasa - Ingeniería de Aplicaciones, S.A.- http://idasa.com/en/ - manufacturer of industrial robots

INALI - http://www.inali.com/ - robotic automation solutions provider

INSER ROBÓTICA S.A. - http://www.inser-robotica.com/ - robotics automation solutions provider

it Robotics - http://www.it-robotics.com/ - robotic automation solutions provider

JKE Robotics - http://www.jkerobotics.com/ - provider of accessories and components for industrial robotics systems

Macco Robotics - https://www.maccorobotics.com/ - developer of humanoid robots

Marsi Bionics - http://www.marsibionics.com/ - robotic exoskeleton manufacturer

PAL Robotics - http://pal-robotics.com/en/home/ - humanoid, robot base and stock robot manufacturer for organizations and research

Robotics Lab - http://roboticslab.uc3m.es/roboticslab/ - developer of various research robots

Robotnik Automation S.L.L. - http://www.robotnik.eu/ - robotics platform developer

SADAKO TECHNOLOGIES - http://www.sadako.es - developer of vision systems for robotics serving municipal solid waste industry

Technaid S.L. - http://www.technaid.com/ - technology developer for biomechanics, rehabilitation, motion analysis, virtual reality, robotics

UAV Navigation - http://www.uavnavigation.com/ - autopilots and UAV parts manufacturer

Wake Engineering - http://www.wake-eng.com/ - drone manufacturer

VISIONA - Visiona Control Industrial S.L - http://www.visionasl.com/ - artificial vision systems developer

Sweden

Bioservo Technologies AB - http://bioservo.com/ - developer of wearable technologies

Bitcraze AB - https://www.bitcraze.io/ - quadcopter developer

Furhat Robotics AB - http://www.furhatrobotics.com/ - developer of robots for human interaction

Hobbytra - http://www.hobbytra.se/ - drone manufacturer

IBC Robotics - http://www.ibcrobotics.se/ - container cleaning robots

Integrum AB - http://integrum.se/ - robotic prosthesis developer

Intuitive Aerial AB - http://www.intuitiveaerial.com/ - drone manufacturer

New Innovation Management - http://new-innovation.se/ - robotic automation solutions provider

Switzerland

ABB – http://www.abb.com – industrial robots manufacturer and installer

Aeroscout GmbH - http://www.aeroscout.ch/ - drone manufacturer

BlueBotics SA - http://www.bluebotics.com/ - developer of various mobile robots

F&P Robotics AG - http://www.fp-robotics.com/en/ - manufacturer of collaborative robots

Hocoma AG - https://www.hocoma.com/ - manufacturer of physical therapy robots

Insider Modellbau GmbH - http://www.insider-modellbau.ch/ - drone manufacturer

K-Team Corporation - http://www.k-team.com/ - mobile robot developer for advanced education and research

MABI AG - Robotic - http://mabi-robotic.com/en/ - manufacturer of industrial robot arms, robot cells, AGVs

MT Robot AG - http://www.mt-robot.com/ - manufacturer of robots for healthcare industry

noonee AG - http://www.noonee.com/ - wearable ergonomic mechatronic devices manufacturer

Pix4D SA - http://www.pix4d.com/ - drone imaging software developer

Rapyuta Robotics Co Ltd. - http://rapyuta-robotics.com/ - developer of cloud robotics systems

Sensars - http://www.sensars.com/ - robotic limb manufacturer

SenseFly - https://www.sensefly.com/home.html - drone manufacturer

Staubli International AG - http://www.staubli.com/ - industrial robot manufacturer

Swisslog Holding AG - http://www.swisslog.com/en - developer of logistics robots for warehouses and hospitals

Taiwan

BeRobot - GeStream Technology Inc. - http://www.berobot.com - humanoid robot toy manufacturer

GAUI - TAI SHIH HOBBY CORPORATION (TSHobby GAUI)- http://www.gaui.com.tw/ - drone manufacturer

HOBOT Technology Inc. - http://www.hobot.com.tw - window cleaning robot manufacturer

MATSUTEK Co., LTD. - http://www.matsutek.com.tw/en/ - manufacturer of vacuum cleaner robots

MKS Servo Tech - http://mks-servo.com.tw/ - RC model servos, gear boxes and electronic control equipment

TT Robotics - http://www.ttrobotix.com/ - manufacturer of hobby aerial, ground, sea drones, robot kits

UAVER - http://www.uaver.com/ - drone manufacturer

XYZrobot - XYZprinting, Inc. - http://www.xyzrobot.com - manufacturer of educational robot kits

Thailand

¡Drones & Co. - http://store.jdrones.com/default.asp - drone manufacturer, distributor

MedicoRobotics Co., Ltd. - http://medicorobotics.com/index.php/en/ - robotic exoskeleton developer

Turkey

AKINSOFT - http://www.akinrobotics.com/ - humanoid robot developer

ALTINAY- http://www.altinay.com/ - robotics automation systems provider

Arge Mekanik - http://www.argemekanik.com/ - manufacturer of industrial robots, automation solutions provider

ASELSAN A.Ş. - http://www.aselsan.com.tr/en-us/Pages/default.aspx - UAV and various robotics systems manufacturer for defense systems

BAMA Teknoloji - http://www.bamateknoloji.com/en/ - developer of robotic rehabilitation systems

BAYKAR Makina - http://baykarmakina.com/ - drone manufacturer

Elektroland Defence - http://www.elektrolanddefence.com/ - mobile robots manufacturer for security and defense sector

Milvus Robotics - http://milvusrobotics.com/tr - manufacturer of mobile robotics platforms

Robotel Turkiye - http://www.robotel.org/ - developer of robotic hands

Robotpark - http://www.robotpark.com.tr/ - robot and electronics parts seller

RTS Robotics - http://www.rtsotomasyon.com/ - robotics automation solutions

Tara Robotik Otomasyon - http://www.tararobotik.com/tr/ - robotics automation services

Türk Havacılık ve Uzay Sanayii A.Ş - https://www.tai.com.tr/tr - UAS manufacturer for defense systems

United Kingdom

Active Robots Ltd. http://www.active-robots.com/ - robot kits and parts seller

Applied Machine Intelligence Ltd. - http://www.appliedmachineintelligence.co.uk/ - developer of educational and personal robots

Arm Limited - https://www.arm.com/ - semiconductor and software developer

Autotech Robotics Limited - http://www.autotech-robotics.com/ - automated manufacturing solutions

Blue Bear Systems Research Limited - http://www.bbsr.co.uk/ - supplier of research and product based solutions in the field of unmanned flight, avionics and sensor payloads

Cambridge Medical Robotics Limited - http://cmedrobotics.com/ - surgical robots manufacturer

CKF Systems Limited - http://www.ckf.co.uk/ - industrial automation solutions developer

DeepMind Technologies Limited - https://deepmind.com/ - developer of AI systems

FiveAI Inc. - http://www.five.ai/ - developer of autonomous vehicle software

Flymo - Husqvarna AB - http://www.flymo.com/uk/ - manufacturer of robotic lawnmowers

Hagen Automation Ltd - https://hagenautomation.com/ - industrial robots manufacturer

Moley Robotics - http://www.moley.com/ - manufacturer of robotic kitchens

Omnicell Ltd. - http://www.omnicell.co.uk/ - pharmaceutical robot manufacturer

Open Bionics - https://www.openbionics.com/ - bionic hands manufacturer

Oxford Technical Solutions - http://www.oxts.com/ - inertial navigation and GPS/GNSS manufacturer

Oxford Technologies, Ltd. - http://www.oxfordtechnologies.co.uk/ - remote handling robots

QuestUAV - http://www.questuav.com/ - drone manufacturer

Raspberry Pi Foundation - https://www.raspberrypi.org/ - open source platform for hobby robotics

Robomow - https://robomow.com/en-GB/ - manufacturer of robotic mowers

RoboSavvy Ltd. - https://www.robosavvy.com - hobby and consumer robots & parts seller

Robot Electronics - Devantech Limited - https://www.robot-electronics.co.uk/ - electronics and robot products and parts manufacturer and seller

Rsl Steeper - http://rslsteeper.com/ - robotic prosthetics manufacturer

Shadow Robot Company Ltd. - http://www.shadowrobot.com/ - developer of robotic hands

Soil Machine Dynamics Ltd. - http://smd.co.uk/ - developer of remote controlled robots for disaster areas, undersea, and various other applications

ST Robotics - Sierra Tango Robotics - https://www.strobotics.com/ - gripper manufacturer

Zettlex UK Ltd. - http://www.zettlex.com/ - drone manufacturer

United States

3D Robotics Inc. - https://3dr.com/ - developer and service provider in UAV technologies

5D Robotics Inc. - http://5drobotics.com/ - autonomous navigation systems developer for drones and autonomous vehicles

Acroname Inc. – http://www.acroname.com – parts manufacturer for embedded robotic, automation and control systems

Action Drone Inc. - http://www.actiondroneusa.com/ - drone manufacturer

Actuonix Motion Devices Inc. - http://www.actuonix.com/ - actuator and servo manufacturer

Aerovel - http://aerovelco.com/ - UAS manufacturer

Aethon Inc. - http://www.aethon.com/ - manufacturer of hospital robots

Agile Sensor Technologies - http://www.agilesensors.com/ - drone manufacturer

Agility Robotics - http://www.agilityrobotics.com/ - manufacturer of legged robots for various applications

Aido - InGen Dynamics Inc. - http://www.aidorobot.com - manufacturer of home robots

AirCover Integrated Solutions - http://www.aircoversolutions.com/ - drone manufacturer

Amazon Robotics - https://www.amazonrobotics.com/ - fulfillment center automation systems as a subsidiary of Amazon.com Inc.

AMP Robotics - http://amprobotics.com/ - developer of recycling robots

ANKI- https://anki.com/en-us - developer of artificial intelligence for consumer robotics, robot manufacturer

Anybots 2.0 Inc. - http://www.anybots.com/ - telepresence robot manufacturer

APC Propellers Landing Products Inc. - https://www.apcprop.com/ - propeller manufacturer for drones

Applewhite Aero - http://www.applewhiteaero.com/ - drone manufacturer

Aqua Products Inc. - https://aquaproducts.com/ - manufacturer of pool cleaning robots

Arduino – http://www.arduino.cc – open source platform for hobby robotics and electronics

Aspect Automation - http://www.aspectautomation.com/ - industrial automation solutions developer

ASTROBOTIC - https://www.astrobotic.com/ - developer of robots for space exploration

ATI Industrial Automation Inc. - http://www.ati-ia.com/ - developer of robotic accessories and robot arm tooling

Atlas Robotics - http://atlasrobotics.com/ - maker of entertainment model robots

Auris Surgical Robotics - http://www.aurisrobotics.com/ - surgical robotics manufacturer

Auro Robotics - http://www.auro.ai/ - manufacturer of self driving shuttles

Aurora Flight Sciences - http://www.aurora.aero/ - drone manufacturer

Autel Robotics USA - https://www.autelrobotics.com/ - drone manufacturer

AUTONOMOUS MARINE SYSTEMS INC. - http://www.automarinesys.com/ - developer of autonomous sea drones

ASI - Autonomous Solutions Inc. - http://www.asirobots.com/ - hardware and software developer for autonomous vehicles

Autonomous Tractor Corporation - http://www.autonomoustractor.com/ - manufacturer of autonomous tractors

AxiDraw - Evil Mad Science LLC - http://www.axidraw.com/ - drawing and writing robot developer

Barret Technology LLC - http://barrett.com - developer of robotic arms, hands and mobile manipulators

BeatBots LLC - http://beatbots.net/ - manufacturer of robot toys

BionX Medical Technologies Inc. - http://www.bionxmed.com/ - developer of robotic prosthetics systems

Blue Robotics Inc. - http://www.bluerobotics.com/ - manufacturer of underwater robots

Bluefin Robotics- General Dynamics Mission Systems, Inc. - http://www.bluefinrobotics.com/ - developer of autonomous underwater vehicles

Bobsweep - http://www.bobsweep.com/ - manufacturer of vacuum cleaner robots

Bossa Nova Robotics - http://www.bossanova.com/ - manufacturer of retail robots

Boston Dynamics - http://www.bostondynamics.com/ - maker of quadruped, bipedal, humanoid and other types of robots

Boston Engineering Corporation - http://www.boston-engineering.com/ - engineering solutions provider including robotics, software, motion control, industrial design, wireless, IoT

Brain Corporation - http://www.braincorp.com/ - developer of autonomous navigation technologies

Caddy Trek - FTR Systems Inc. - https://www.caddytrek.com/ - manufacturer of autonomous golf carts

Canvas Technology - http://www.canvas.technology/ - developer of autonomous mobile robots for factories and warehouses

Carbon by Design - http://www.carbonbydesign.com/ - drone manufacturer

Carbon Robotics - http://www.carbon.ai/ - developer of industrial robotic arm

Case IH - https://www.caseih.com/ - autonomous tractor manufacturer

Cognitoys - Elemental Path Inc. - https://cognitoys.com/ - robotic toys manufacturer

Construction Robotics - http://www.construction-robotics.com/ - developer of construction robots

Corindus Inc. - http://www.corindus.com - surgical robot manufacturer

Cruise - https://www.getcruise.com/ - developer of driverless car technologies

CrustCrawler Robotics - http://www.crustcrawler.com/ - developer of robotic arms, walking robots, grippers

Custom Entertainment Solutions Inc. - http://animatronicrobotics.com/ - designer of animation robots

Cyberworks Robotics Inc. - http://cyberworksrobotics.com/ - autonomous mobile engineering, developer of autonomous wheelchairs and vacuum cleaner robots

CyPhy Works, Inc- http://www.cyphyworks.com/ - drone manufacturer

DEKA Research & Development Corp. - http://www.dekaresearch.com/ - developer of robotic arms, wheelchairs, self balancing transporters

Delta Tau Data Systems - http://www.deltatau.com/ - developer of motion controllers

DENSO Robotics - http://densorobotics.com/world/ - manufacturer of industrial robots

Digi-Key Electronics - http://www.digikey.com/ - electronics parts seller

Discovery Robotics - http://www.discovery-robotics.com/ - developer of commercial floor cleaning robots

Double Robotics, Inc. - https://www.doublerobotics.com/ - manufacturer of telepresence robots

DPI UAV Systems - DRAGONFLY PICTURES, INC - http://www.dragonflypictures.com/ - drone manufacturer

Draganfly Innovations Inc- http://www.draganfly.com/ - drone manufacturer

DU-BRO RC - http://www.dubro.com/ - RC products manufacturer

Eagle Tree Systems, LLC - http://www.eagletreesystems.com - RC products manufacturer

Ecovacs Robotics, Inc. - http://ecovacsrobotics.com/ - manufacturer of robotic home appliances

Ekso Bionics - http://eksobionics.com/ - robotic exoskeleton developer

Energid Technologies Corporation- http://www.energid.com/ - manufacturer of industrial robots

Equipois, LLC - http://www.equipoisinc.com/ - developer of mechanical arms

Ergopedia - http://ergopedia.com - developer of robotic kits and educational robots

FarmBot Inc. - https://farmbot.io/ - manufacturer of farm and garden robots

Fendt - AGCO Corporation - http://www.fendt.com/ - mobile agricultural robot swarms, autonomous tractor developer

Fetch Robotics Inc. - http://fetchrobotics.com/ - robotics systems for logistics industry

FLINT HILLS SOLUTIONS - http://www.fhsllc.com/ - UAS manufacturer and services provider

FLIR Systems Inc. - http://www.flir.eu/home/ - advanced imaging and sensing systems developer

FMC Technologies - http://www.fmctechnologies.com - manufacturer of remotely operated vehicles (ROVs) and manipulator arms

Franklin Robotics LLC - http://www.franklinrobotics.com/ - developer of garden robots

Gamma 2 Robotics - http://www.gamma2robotics.com/ - manufacturer of security robots

GetFPV LLC - http://www.getfpv.com/ - drone parts manufacturer

Ghost Robotics LLC - http://www.ghostrobotics.io/ - developer of gearless legged robots

GoPro Inc. - http://gopro.com/ - camera manufacturer for drones and high action systems

Greensea Systems Inc. - https://greenseainc.com/ - developer of navigation and automation components

Grillbots - https://grillbots.com/ - grill cleaning robot manufacturer

Hacker Motor Distribution Co. - http://www.hackermotorusa.com - motor, servo, ESC, fan manufacturer

Hansen Medical - http://www.hansenmedical.com/us/en - developer of medical robots

Harvest Automation - http://www.public.harvestai.com/ - developer of autonomous robots for agriculture

Helimax - Hobbico Inc. - http://www.helimaxrc.com/ - multirotor manufacturer

HEXBUG - Innovation First Labs, Inc. - https://www.hexbug.com/ - robot toys manufacturer

Hexo+ - http://www.hexoplus.com/ - drone manufacturer

Hirebotics - http://www.hirebotics.com/ - provider of robotics automation services

Hobbico Inc. - http://www.hobbico.com- manufacturer, distributor and seller of RC & general electronics hobby products

Hobby Express - http://www.hobbyexpress.com/ - manufacturer, distributor and seller of RC & general electronics hobby products and parts

Hobbyking - http://www.hobbyking.com/ - manufacturer, distributor and seller of RC & general electronics hobby products and parts

Honeybee Robotics Ltd. - https://www.honeybeerobotics.com/ - developer of robots for defense, medical, space and mining industries

Horizon Hobby – http://www.horizonhobby.com – manufacturer, distributor and seller of RC & general electronics hobby products and parts

Hylio Inc. - http://www.hyl.io/ - drone manufacturer

IAM Robotics - https://www.iamrobotics.com/ - manufacturer of autonomous robots for warehouses

Indego - Parker Hannifin Corp. http://www.indego.com/indego/en/home -exoskeleton manufacturer

INOVA Drone - http://www.inovadrone.com/ - drone manufacturer

Innovation Associates - http://innovat.com/ - manufacturer of pharmaceutical robotics

Intellibot Robotics - http://www.intellibotrobotics.com/ - manufacturer of cleaning robots

Intuitive Surgical, Inc. - http://www.intuitivesurgical.com/ - manufacturer of surgical robots

InVia Robotics Inc. - http://www.inviarobotics.com/ - fulfillment center robots manufacturer

iRobot Corporation – http://www.irobot.com – manufacturer of consumer robots including vacuum cleaner and mopping robots

Iron Ridge Engineering LLC - http://www.ironridgeuas.com/ - UAS manufacturer

Jameco Electronics - http://www.jameco.com/ - hobby and consumer robots, electronics products & parts seller

Jibo Inc. - https://www.jibo.com/ - manufacturer of personal entertainment robot

John Deere - Deere & Company - https://www.deere.com/ - autonomous tractor manufacturer

JOY FOR ALL - Hasbro - https://joyforall.hasbro.com/en-us - robotic companion pet manufacturer

JR Automation Technologies LLC - http://www.jrauto.com/ - developer of automation solutions for a variety of industries

Kespry - http://kespry.com/ - drone manufacturer, drone services provider

Kimera Systems - http://kimera.ai/ - developer of artificial intelligence technologies

Knightscope Inc. - http://www.knightscope.com/ - manufacturer of security robots

Leptron Unmanned Aircraft Systems Inc. - http://www.leptron.com/ - UAS manufacturer

Liquid Robotics Inc. - The Boeing Company - https://www.liquid-robotics.com/ - manufacturer of autonomous sea robots

Litter Robot - Automated Pet Care Products, Inc. - https://www.litter-robot.com/ - manufacturer of self cleaning litter box robots for pets

Locus Robotics - http://www.locusrobotics.com/ - manufacturer of warehouse robots

LRP America - http://www.lrp-americastore.com/ - drone manufacturer seller

Lynxmotion - Robotshop Inc. – http://www.lynxmotion.com – manufacturer of hobby and research robot kits and parts

Mars Parachutes - http://www.marsparachutes.com/ - drone parachutes manufacturer

Martin UAV - http://martinuav.com/ - drone manufacturer and drone services provider

MathWorks - The MathWorks Inc. - https://www.mathworks.com/ - developer of mathematical computing software for engineers and scientists, for different fields in AI and automation

Mayfield Robotics - http://www.mayfieldrobotics.com/ - manufacturer of home robots

Maytronics - http://www.maytronics.com/ - pool cleaning robot manufacturer

Meccano - http://www.meccano.com/ - manufacturer of robot kits

Medrobotics Corporation - http://medrobotics.com/ - surgical robots manufacturer

Metapo Inc. - http://www.metapo.com/ - manufacturer of robotic vacuum cleaners

Microbot Medical Inc. - http://www.microbotmedical.com/ - developer of micro-robotic medical technologies

MINDS-i Inc. - http://mindsieducation.com/ - developer of educational robot kits

MODBOT - https://www.modbot.com/ - manufacturer of modular robot building platforms

Modular Robotics Incorporated - http://www.modrobotics.com/ - developer of modular toy robots

Motorika USA Inc. - http://motorika.com/ - developer of rehabilitation robots

Mouser Electronics Inc. - TTI and Berkshire Hathaway - http://www.mouser.com - electronic components seller

Myomo Inc. - http://www.myomo.com/ - manufacturer of medical robots for physical rehabilitation

Naturaldrones USA LLC - http://www.naturaldrones.com/ - drone manufacturer

Neato Robotics Inc. - https://www.neatorobotics.com/ - robot vacuum cleaner manufacturer

Neurala Inc. - http://www.neurala.com/ - developer of artificial intelligence for various types of robots

NewBotic Corporation - http://www.newbotic.com/ - robotic systems integrator

Northrop Grumman Corportation - http://www.northropgrumman.com – provider of systems, products, solutions for defense and commercial sectors for a variety of applications

nuTonomy - http://nutonomy.com/ - self driving systems developer

Ocean Aero - http://www.oceanaero.us/ - manufacturer of sea drones

Oceanic Robotic Systems - http://gotransat.com/ - developer of autonomous transatlantic boat

Olympus Controls - http://www.olympus-controls.com/ - engineering services that specializes in automation

Omron Adept MobileRobots LLC - http://www.mobilerobots.com/Mobile_Robots.aspx - mobile robot systems for research and education

ONAGOfly - http://www.onagofly.com/ - drone manufacturer

Osaro - http://www.osaro.com/ - artificial intelligence, and machine intelligence software developer for industrial, consumer robots, autonomous vehicles, drones and other systems

OWI Inc. - http://www.owirobot.com/ - educational robot kits manufacturer

Ozobot & Evollve, Inc. - http://ozobot.com/ - robotic toy maker

Parallax Inc. – http://www.parallax.com – electronics and robot products and parts manufacturer and seller

Parker Hannifin Corp. - http://www.parker.com - motion and control technology products and solutions developer for industrial and aerospace markets

PARO Robots U.S., Inc. - http://www.parorobots.com/ - interactive robot developer

Perceptron Inc. - http://perceptron.com/ - metrology equipment and solutions provider

Petnet Inc. - http://www.petnet.io/ - robotic pet feeder manufacturer

Phomatix LLC - http://www.phomatix.com/ - maker of social and mapping robots for photo and video capturing

PlySync - https://polysync.io/ - developer of middleware platforms for autonomous vehicles

Pololu Corporation – http://www.pololu.com – electronics and robot products and parts manufacturer and seller

Quantum - http://www.quanum-rc.com/index.php - drone and drone parts manufacturer and seller

RailPod Inc. - http://rail-pod.com/ - developer of railroad inspection robots

RE2 Inc. - http://www.resquared.com/ - developer of intelligent modular manipulation systems and mobile robotics technologies

ReconRobotics - http://www.reconrobotics.com/ - developer of tactical, micro-robot systems for defense and law enforcement sectors

Redwood Robotics - https://www.redwood.com/robotics/ - business automation solutions provider

RedZone Robotics - http://www.redzone.com/ - wastewater inspection robots for municipalities, contractors, and engineering companies

Reis Robotics - http://www.reisroboticsusa.com/ - robotics system integration provider

Restoration Robotics Inc. - http://restorationrobotics.com/ - developer hair surgery robots

Rethink Robotics - http://www.rethinkrobotics.com/ - developer of adaptive robots for manufacturing

Rewalk Robotics - http://rewalk.com/ - robotic exoskeleton manufacturer

RightHand Robotics Inc. - http://www.righthandrobotics.com/ - industrial robot manufacturer

Rise Robotics - http://www.riserobotics.com/ - mechanical components developer for robotic exosuits

RoadNarrows Robotics - https://roadnarrows.com/ - developer of robot platforms, vision modules, 3D vision, sensor networks, open-source platform software

Robai Corporation - http://www.robai.com/ - lightweight dextrous robot manufacturer

Robomatter Inc. - https://robomatter.com/ - educational robotics software developer

Roboteam - http://www.robo-team.com/ - manufacturer of tactical ground robotic systems

RobotGeek - http://www.robotgeek.com/ - hobby robot platforms and kits provider

Robotics Inventions - http://www.roboticsind.com/ - developer of vision systems, commercial, security, personal and swarm robots

ROBOTIS - http://en.robotis.com/index/ - manufacturer of robotic platforms, educational kits, humanoid robots, software, actuators

RobotLAB Inc. - http://www.robotlab.com/ - developer of educational robot kits

Rogue Robotics - https://roguerobotics.com/ - electronics and robot products and parts manufacturer and seller

SapientX Inc. - http://www.sapientx.com/ - developer of conversational AI software

Sarcos Corp. - http://www.sarcos.com/ - developer of dextrous robotics products for use in unstructured environments

Savioke - http://www.savioke.com/ - developer of autonomous robot helpers for services industry

Scion UAS LLC - http://www.scionuas.com/ - drone manufacturer

Seeed Development Limited - http://www.seeedstudio.com/index.html - electronic components manufacturer

Seegrid Corporation - https://seegrid.com/ - manufacturer of VGVs for manufacturing and industrial facilities

Segway - http://www.segway.com - manufacturer of two wheeled self balancing robots

Sewbo Inc. - http://www.sewbo.com/ - developer of sewing robots

Sharp Electronics Corporation - http://www.sharpintellos.com/ - outdoor security, surveillance and maintenance robot manufacturer

SHOTOVER - http://shotover.com/ - drone manufacturer, drone services

Silent Falcon - http://www.silentfalconuas.com/ - UAS manufacturer

SIMBE ROBOTICS INC. - http://www.simberobotics.com/ - developer of autonomous robots for supermarkets

Skygen Aviation - http://www.skygenaviation.com/ - drone manufacturer, drone services

Skyspecs - http://www.skyspecs.com/ - drone manufacturer, drone services

Smartpool LLC - http://www.smartpool.com - robotic pool cleaner manufacturer

Smith & Nephew - http://www.smith-nephew.com - developer of surgical robots

SMP Robotics Systems Corp. - http://www.smprobotics.com - UGV manufacturer

Solar Breeze - Solar Pool Technologies, Inc - https://solar-breeze.com/ - robotic solar pool cleaner manufacturer

Soft Robotics Inc. - http://www.softroboticsinc.com/ - robotic gripper manufacturer

Sparkfun – http://www.sparkfun.com – electronics and robot products and parts manufacturer and seller

Spectracom Corp. - http://spectracom.com/ - unmanned and automated solutions provider for defense, industrial, automative sectors

Sphero - http://www.sphero.com/ - manufacturer of toy and educational robots

Spinmaster Ltd. - http://www.spinmaster.com - robot toy manufacturer

SpringActive Inc. - http://www.springactive.com/ - robotic prostheses manufacturer

Square Robot Inc. - http://www.squarerobots.com/ - developer of autonomous solutions for inspection systems

Stanley Innovation - http://stanleyinnovation.com/ - customized robotics solutions for businesses and researchers

Suitable Technologies Inc. - https://www.suitabletech.com/ - manufacturer of telepresence robots

suitX - US Bionics Inc. - http://www.suitx.com/ - robotic exoskeleton manufacturer

SuperDroid Robots - http://superdroidrobots.com/ - manufacturer of tactical robots, robot kits, SWAT robots

Swift Navigation Inc. - https://www.swiftnav.com/ - navigation solutions for unmanned vehicles

SynTouch Inc. - https://www.syntouchinc.com/ - developer of tactile sensing for industrial, defense and healthcare applications

Teradyne - http://www.teradyne.com/ - supplier of automation equipment for test and industrial applications

Texas Instruments Incorporated - http://www.ti.com/ - electronic components manufacturer and seller

TorRobotics - http://torrobotics.com/ - drone manufacturer

TORC ROBOTICS - http://torcrobotics.com/ - developer of remote and autonomous systems for vehicles

Touch Bionics Inc.. - http://www.touchbionics.com/ - maker of robotic prosthetic limbs

Transcend - http://transcendrobotics.com/ - developer of UGVs for various applications

TransEnterix Surgical Inc. - http://www.transenterix.com/ - surgery robots manufacturer

Traxxas - http://www.traxxas.com/ - RC products manufacturer and seller

Trossen Robotics - http://www.trossenrobotics.com/ - hobby robots, robot kits and parts seller

UAV Factory - http://www.uavfactory.com/ - UAV manufacturer

UAV Solutions Inc. - http://www.uavsolutions.com/ - UAV manufacturer

UAVX LLC - http://www.uavx.com/ - drone remote control systems developer

ULC Robotics - http://ulcrobotics.com/ - developer of robotic inspection systems for a variety of applications

Vecna - https://www.vecna.com/ - developer of telepresence, autonomous warehouse and hospital robots

Verb Surgical Inc. - http://www.verbsurgical.com/ - developer of surgical robots

Vertical Partners West LLC - http://www.vpwllc.com/ - drone manufacturer

Vex Robotics - Innovation First International Inc. - http://www.vexrobotics.com/ - educational and hobby robotics kits manufacturer

VGo Communications Inc. - http://www.vgocom.com - manufacturer of telepresence robots

Video Aerial Systems LLC - http://videoaerialsystems.com/ - drone and drone components manufacturer

Vincross - https://www.vincross.com/ - toy robot manufacturer

Vision Aerial LLC - http://www.visionaerial.com/ - drone manufacturer

WHEEME - Dreambots LTD. - http://wheeme.com/ - massage robot manufacturer

WiBotic - http://www.wibotic.com/ - wireless power and battery intelligence solutions to charge land sea or aerial robot systems

Willow Garage Inc. - http://www.willowgarage.com/ - developer of hardware and open source software for personal robotics applications.

WinSystems Inc. - https://www.winsystems.com/ - embedded systems manufacturer

Wonder Workshop Inc. - https://www.makewonder.com/ - robot toy developer

Wynright Corporation - http://www.wynright.com/ - intelligent material handling systems for industrial applications

Xenex Disinfection Services - https://www.xenex.com/ - hospital cleaning robot developer

Yaskawa Motoman - Yaskawa America Inc. - https://www.motoman.com/ - industrial, educational, laboratory robots manufacturer, automation solutions developer

Zargos Robotics - http://www.zagrosrobotics.com/shop/ - robot products and parts seller

ZiroUI Inc. - http://ziro.io/ - robot kit developer

Zodiac Pool Systems Inc. - http://www.zodiacpoolsystems.com/en - robotic pool cleaner manufacturer

Zyrobotics LLC - http://zyrobotics.com/ - robot toy manufacturer

CHAPTER 14

VARIOUS SOURCES

Events

Conferences & Expositions

Age of Drones - http://www.ageofdrones-expo.com/

Airborne ISR - http://www.smi-online.co.uk/defence/uk/airborne-isr

Automatica - http://automatica-munich.com

AUVSI Events - http://www.auvsi.org/events1aa/events

CES - http://www.ces.tech/

Commercial UAV Expo - http://www.expouav.com/

Drone World Expo - http://www.droneworldexpo.com/

Global Robot Expo - http://www.globalrobotexpo.com/en/

International Conference on Social Robotics - http://icsoro.org/

IEEE Events - http://www.ieee-ras.org/about-ras/ras-calendar/upcoming-ras-events

Ind. Automation N.America-https://industrialtechnology.events/industrial-automation-north-america/

Interdrone - http://www.interdrone.com/

International Drone - http://www.internationaldroneday.com/

International Drone Expo (IDE). http://internationaldroneexpo.com/

Middle East Robotic Process Automation - https://roboticprocessautomation.iqpc.ae/

Next Gen Automative Robotics - http://www.nextgenautomotiverobotics.com/

RIA Events - https://www.robotics.org/events.cfm

Robo City - http://www.robocity2030.org/

Robobusiness - http://robobusiness.com/

Robouniverse - http://robouniverse.com/

RPA & AI - https://www.rpaandaisummit.com/

Skytech events - http://www.skytechevent.com/

UAS Summit & Expo - http://www.theuassummit.com

UAV Tech Europe - http://www.smi-online.co.uk/defence/europe/UAV-Technology-Eastern-Europe

UK Drone Show - http://www.ukdroneshow.com/

Competitions

Botball - http://www.botball.org/

Drone Racing League - http://thedroneracingleague.com/

Drone Racing World - http://droneracingworld.com/

Eurobot - http://www.eurobot.org/

FIRST - https://www.firstinspires.org/

NASA Space Robotics Challenge, NASA Robotics Initiative - https://www.nasa.gov/directorates/spacetech/centennial_challenges/space_robotics/index.html

https://www.nasa.gov/robotics/index.html

Robocup - http://www.robocup.org

Robogames - http://robogames.net/index.php

Robonation - http://www.robonation.org/

Hobby Robotics & DIY Electronics Sites

https://www.americasgreatestmakers.com/

http://www.diyhacking.com

http://www.firstinspires.org/

http://www.fischertechnik.de

http://hackaday.com/

http://www.instructables.com

http://makezine.com/

https://www.raspberrypi.org/

http://www.robotroom.com

http://www.societyofrobots.com

http://www.vexrobotics.com/

Organizations (USA Only)

Association for Advancing Automation - http://www.a3automate.org/

Association for Advancement of Artificial Intelligence (AAAI) - http://www.aaai.org/home.html

Association for Unmanned Vehicle Systems International (AUVSI) - http://www.auvsi.org/home

Global Association for Vision Information - http://www.visiononline.org/

IEEE Robotics & Automation Society - http://www.ieee-ras.org/

International Federation of Robotics (IFR) - http://ifr.org/

Motion Control Online - http://www.motioncontrolonline.org/

Multi Robot Systems - http://multirobotsystems.org/

NASA - http://www.nasa.gov/topics/technology/robotics/index.html

Open Source Robotics Foundation - http://www.osrfoundation.org

Robotics Industries Association (RIA) - http://www.robotics.org/

News Sources

Automation.com - http://www.automation.com/

AZoRobotics - http://www.azorobotics.com/

Drone Flyers - http://www.droneflyers.com/

Droneblog - http://droneblog.com/

Dronestagram - http://www.dronestagr.am/

Gizmodo - http://gizmodo.com/tag/robots

IEEE Spectrum - http://spectrum.ieee.org/blog/automaton

IEEE Spectrum - http://spectrum.ieee.org/robotics

Live Science - http://www.livescience.com/topics/robots

Naval Drones - http://www.navaldrones.com/

New Scientist - https://www.newscientist.com/article-topic/robots/

Phys.org - http://phys.org/technology-news/robotics/

Popular Mechanics - http://www.popularmechanics.com/technology/robots/

Robo Daily - http://www.robodaily.com/

Robo Hub - http://robohub.org/

Robot Magazine – http://www.botmag.com

Robot Rabbi – http://www.robotrabbi.com

Robotic Magazine – http://www.roboticmagazine.com

Robotics and Automation News - http://roboticsandautomationnews.com/

Robotics Business Review - http://www.roboticsbusinessreview.com

Robotics News - http://robotics.news/

Robotics Tomorrow - http://www.roboticstomorrow.com/

Robotics Trends – http://www.roboticstrends.com

Rotor Drone Magazine - http://www.rotordronemag.com/

Science Daily - https://www.sciencedaily.com/news/computers_math/robotics/

Science Robotics - http://robotics.sciencemag.org/

SUAS News - http://www.suasnews.com/

That Drone Show - http://www.thatdroneshow.com/

The Drone Files - http://www.thedronefiles.net/

The Robot Report - https://www.therobotreport.com/

The UAV Digest - http://www.theuavdigest.com/

Thomasnet - http://news.thomasnet.com/news/robotics

UAS Magazine - http://www.uasmagazine.com/

Unmanned Systems Technology - http://www.unmannedsystemstechnology.com/

Robot, Drone and Electronics Discussion Forums

Robotic Magazine - http://www.roboticmagazine.com/forums/

All About Circuits http://www.allaboutcircuits.com/

Arduino - https://forum.arduino.cc/

ArduPilot - http://www.ardupilot.com/forum/

DIY Drones - http://diydrones.com/

Drone Flyers - http://www.droneflyers.com/talk/

Dronestagram - http://www.dronestagr.am/forums/

Engineers Garage - https://www.engineersgarage.com/forum

Fighting Robots UK - http://www.fightingrobots.co.uk/forum

Multi Rotor Forums - http://www.multirotorforums.com/

Multi Wii - http://www.multiwii.com/forum/

Parallax - http://forums.parallax.com/categories/robotics

Pololu - https://forum.pololu.com/

Rasberry Pi - https://www.raspberrypi.org/forums/

RC Groups - http://www.rcgroups.com/forums/index.php

Robot Forum - http://www.robot-forum.com/robotforum/

Robot Reviews - http://www.robotreviews.com/chat/

Robot Studio https://forums.robotstudio.com/

RobotC - http://www.robotc.net/forums/

Robotiq - http://dof.robotiq.com/

Robotshop - http://www.robotshop.com/forum/

ROS - http://answers.ros.org/questions/

Society of Robots - http://www.societyofrobots.com/robotforum/

Sparc - http://sparc.tools/forum/

Sparkfun - https://forum.sparkfun.com/

Traxxas - https://traxxas.com/forums

Trossen Robotics - http://forums.trossenrobotics.com/content.php

Made in the USA
San Bernardino, CA
05 October 2017